METHODS

FOR THE

COMPUTATION FROM DIAGRAMS

OF

PRELIMINARY AND FINAL ESTIMATES

OF

RAILWAY EARTHWORK,

WITH DIAGRAMS

GIVING QUANTITIES ON INSPECTION TO THE NEAREST CUBIC YARD, FOR BOTH REGULAR AND IRREGULAR SECTIONS, DIRECT FROM ORDINARY FIELD-NOTES.

BY

ARTHUR M. WELLINGTON,

CIVIL ENGINEER.

PART I.—TEXT.

NEW YORK:
D. APPLETON AND COMPANY,
549 & 551 BROADWAY.
1875.

PREFACE.

THE nearest approach to this somewhat novel method of computing earthwork is found in the methods often used for graphically computing bridge-strains, and occasionally for preliminary estimates; both being based on Analytical Geometry. But there is the important difference in this case that no *construction* is required, or scaling of distances, the quantities being merely read off, as if from a table. Inaccuracy and delay are thus avoided, and the method becomes adapted to numerous and exact computations.

The best description, in fact, which can be made of the diagrams is, that they are a series of CONDENSED TABLES; the only difference being that quantities are read off from lines instead of Arabic numerals, which, with a little practice, is equally convenient. The diagrams might be replaced by tables of the ordinary form, except for three reasons:

First. The tables would be of vast extent; larger, indeed, than all those which have ever been made for earthwork computation put together.

Second. It would be impracticable to secure accuracy in such extensive tables; whereas, in constructing a diagram, any error of importance becomes immediately evident.

Third. Even if the tables were accurately constructed, they would be too bulky for practical use.

These objections are self-evident, if the first be granted. To illustrate that point, let us take a single diagram, Plate I. It extends to about 29 feet centre-heights, with a range in horizontal dimensions of 32 feet. Then to tabulate it, to tenths of each dimension, would require $290 \times 320 =$ 89,600 quantities, equal to 155 pages of Trautwine's tables, or ten times as extensive a series of tables as all those given in that volume, which are for 12 different road-beds, and extend to 60 feet centre-heights instead of 30 feet. Moreover, every *point* on Plate I. represents a tabular quantity, and this fact enables the range of the diagram to be readily quadrupled, which cannot be conveniently done with tables.

This statement sufficiently illustrates the peculiar advantage of diagrams; that their conciseness and ease of construction enable results to be reached which are otherwise quite unattainable. The only plausible objection which can be made *against* their use, is a lack of decimal precision; and it is so frequently urged as to require some notice, though quite unjustified by facts.

The diagrams for construction estimates read to the nearest cubic yard, in all cases, or to half-feet of sectional area. Now, the unavoidable irregularities of construction render this limit abundantly accurate even if the field-notes were absolutely correct, and the method of computation perfect. As a matter of fact, they are never so. In the field-work, half-tenths are habitually neglected, and there are other sources of error. In computation, nine-tenths of railway estimates are made by averaging end-areas, which is widely erroneous, and the remaining tenth by the "Method of Level Sections," involving errors greatly in excess of the "error of observation," and tending always to deficiency. In fact, it may safely be asserted that no theoretically perfect earthwork estimate of any extent has ever been made, because the labor ordinarily required is quite incommensurate with the importance of the result. By the aid of diagrams, however, an estimate may be made with a theoretical nicety now rarely if ever attempted; or, if preferred, we may rest content with any degree of approximation; but, in either case, quantities are determined directly from the notes, with little or no computation. When such a result is rendered possible, a minute compensating error may safely be left to take care of itself—although it may prove a fatal objection to that numerous class who condemn "theoretical refinements," but carry out "end-area" solidities, with great care, to decimals of a yard.

It is not supposed that all the diagrams given will be of equal value in practice. The Diagram of Cross-Sections and of Triangular Prisms will probably be of most general convenience; especially the latter, from its very general application to all varieties of earthwork. Diagrams of Cross-Sections for double-track road-beds are not included among the plates, because levels over the road-bed angles are commonly taken with wide road-beds, and the section is then best computed from the Diagram of Triangular Prisms. The Diagrams of Cross-Sections may be applied, however, to all road-beds of the three most common slopes, by the simple process of paragraph 105, so that only a single plate for each side-slope is absolutely required. Only a single diagram for computation by the method of Henck's "Field-book for Engineers" is included, on account of the comparatively rare use of the method. The single diagram given is applicable to all road-beds having 1½ to 1 side-slopes by the process of paragraph 109.

The Diagram of Prismoidal Correction can be used to determine corrections either by the "Method of Level Sections" or by what has been here termed the "Method of Centre-Heights," and it also offers a concise and theoretically perfect method, which is something of a novelty, and is due to Prof. Charles A. Smith, of Washington University. For the benefit of those interested in exact computation, a comparison of these methods is given in Appendix A, not on account of the intrinsic importance of their discrepancies, but to bring out the fact, which the writer regards as of some importance, that the "Method of Level Sections" has an injurious effect on an estimate as compared with a simpler process; and that the same holds true of some elaborate rules for computing irregular earthwork which have been recommended and possibly used. The practical effect of such methods is mainly visible in a very general verdict that they "don't pay," and in the consequent almost universal use of simple "end-area" solidities. The fact that more correct results can be reached with so little labor as to be unobjectionable to every one is at least not generally recognized, and is thought to be here first demonstrated. To reduce this labor to its lowest terms a little table has been added, in Appendix B, for the reason there stated. The simple expedient of determining separate corrections at any time, to be deducted in the sum total, is not known to have been previously suggested, though it greatly simplifies exact computation.

The methods given for preliminary estimates are entirely new, except that Trautwine's diagrams and tables* subserve the same purpose indirectly as the diagram for a uniform surface-slope. They are especially designed for use with an odometer, to enable the volume of an entire cut with either uniform or double surface-slopes to be determined at once. The method of correcting for curvature

* "A New Method of calculating the Cubic Contents of Excavations and Embankments by the Aid of Diagrams." By John C. Trautwine, C. E.

is also new. All the diagrams proper are of course original, but, except as has been mentioned, they are based on methods in general use, and their only purpose is to eliminate multiplication and division from the process, and reduce it to simple addition. They have been constructed with great care, and, it is thought, will prove trustworthy. They are at least absolutely free from important error. It may be proper to add that the numerical illustrations throughout this volume are given as actually taken off from the originals of the plates, and are a fair test as they stand of the accuracy which can be habitually attained with ordinary care. None of them were previously computed, nor have any legitimate errors of observation been corrected.

In the Preliminary Explanation and elsewhere it has been preferred to err on the side of fullness, as will readily be seen. But little familiarity with the ordinary rules for measurement or with mathematics has been assumed, and none whatever with Analytical Geometry.

The writer takes pleasure in acknowledging his indebtedness to his friend Prof. Charles A. Smith, of Washington University, not only for the equation on which the Diagram of Prismoidal Correction is founded (equation 15) and the suggestion to construct a diagram from it, but also for a number of valuable suggestions during an extensive correspondence; including, especially, that of multiplying all areas by the factor $\frac{100}{54}$ to give solidities at once instead of sectional areas.

A. M. W.

New York, *March*, 1874.

CONTENTS.

CHAPTER I.

PRELIMINARY EXPLANATION.

CHAPTER II.

FORMULÆ AND RULES.

ART. I.—COMPUTATION OF END-AREA SOLIDITIES FOR "THREE-LEVEL" SECTIONS.

Diagram of Cross-Sections (Plates I. to VI.).

Diagram of Side-Hill Cross-Sections (Plate VII.).

ART. II.—COMPUTATION OF "THREE-LEVEL SECTIONS" BY THE PRISMOIDAL FORMULA.

Diagram of Prismoidal Correction (Plate IX.).

Diagram for Computation by Henck's Formula (Plate X.).

CHAPTER III.

INDIRECT APPLICATIONS TO OTHER ROAD-BEDS THAN THOSE TO WHICH THE DIAGRAMS DIRECTLY APPLY.

CHAPTER IV.

OFFICE NOTES.

CHAPTER V.

CONSTRUCTION OF DIAGRAMS.

APPENDIX A.

APPENDIX B.

ERRATA.

Page 22—Fig. 8.	Centre height of second section,	for 11.8,	read 9.8
" 137—Head-line to table,		" 8.	" .8
" XV.—Last line of "List of Plates,"		for $1\frac{1}{4}$ to 1,	read $\frac{1}{4}$ to 1
Part II.— " " " "		" "	" "

LIST OF PLATES.

COMPUTATION FROM DIAGRAMS
OF
RAILWAY EARTHWORK.

CHAPTER I.

PRELIMINARY EXPLANATION.

1. Suppose it is required to calculate the area of a number of strips of land, of uniform width but varying length; as in estimating the right of way for a line of railway. It is evident that the number of acres will vary directly as the length of the centre line. In all such cases we can construct a simple figure, from which, if one of those quantities is known, we can obtain, by measurement, the corresponding value of the other.

Fig. 1.

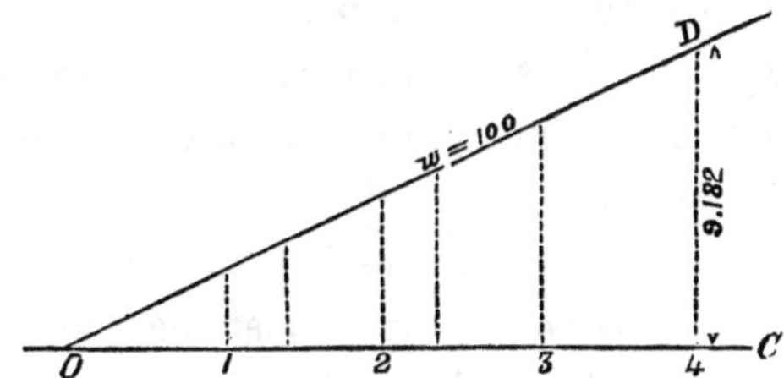

Thus, from any point, O, on the line $O\ C$, Fig. 1, lay off to scale, say 100 feet to an inch, successive lengths of 1, 2, 3, 4 thousand feet. Compute for any one of these lengths, say 4,000 feet, the corresponding area of a strip of land of the given width; e. g., 9.182 acres, for a strip 100 feet wide. Through the point 4, erect a perpendicular to the line *O C* ; and, taking some convenient scale, say five-tenths of an acre to an inch, lay off along it the distance 4 *D*, equal to 9.182 acres. Draw the line *O D*, and erect the verticals 1, 2, 3, etc., parallel with the line 4 *D*.

Now it is evident, by the principle of proportional triangles, that if the line 4 *D* is equal by scale to the number of acres in a strip 4,000 feet long, the parallel lines 1, 2, 3, will be equal by the same scale to the number of acres for a length of 1,000, 2,000, and 3,000 feet respectively; and hence, if any other length, as 1,400 or 2,350 feet, be laid off by scale along the line *O C*, and a perpendicular erected through the point obtained till it cuts the line *O D*, the length of that line also will be equal by scale to the number of acres corresponding to the given length of 1,400 or 2,350 feet. The particular scale employed in no way affects the principle, and by selecting a larger or a smaller one we can obtain the number of acres with any required degree of precision.

2. If these perpendiculars *be previously erected at certain equal distances apart*, and at such small intervals that the position of intermediate lines can be *estimated by the eye without sensible error*, as for example, in the above instance, for each even 10 feet in length of centre line, we shall avoid the necessity of erecting perpendiculars for individual cases, and can scale off the number of acres for a given case at once. Similarly, the use of a scale may be avoided, by drawing *horizontal* lines also so near together that any distance between two lines can be estimated by the eye, with sufficient exactness. We shall thus obtain from Fig. 1 a diagram similar to Fig. 2; the vertical lines serving to

FIG. 2.

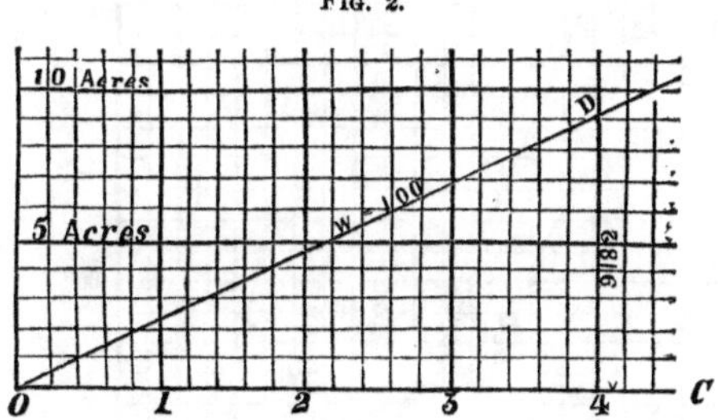

mark off horizontal distances, and the horizontal lines to mark off vertical distances, thereby serving the purpose of a scale. The resemblance of Fig. 2 to ordinary cross-section paper is evident. In practice, the latter is admirably adapted to this purpose, and is always used for it.

3. It is self-evident that inclined lines can be drawn on the same sheet for a number of *different* widths, as 60, 80, and 120 feet, and that they will not interfere with each other, since each will fall in a different place on the sheet. We thus obtain a diagram of which Fig. 3 is a reduction to one-tenth scale, and, when working

10 Acres.
9 "
8 "
7 "
6 "
5 "
4 "
3 "
2 "
1 "
10 Acres
Axis of A.
Axis of L
0
500
1,000
1,500
2,000
2,500
3,000
3,500
4,000
4,500
w=120
w = 100
" 80
" 60
" 40
" 20

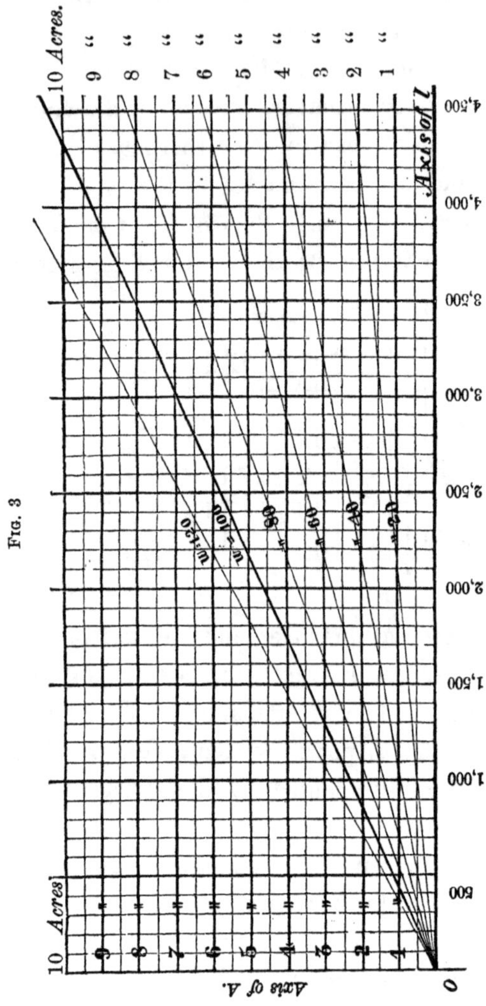

FIG. 3

with any particular width, we can use the line applying to it as if there were no others on the sheet. There is evidently no essential difference between this diagram and that for a single width; for we have only, as it were, constructed several different diagrams on the same sheet, and the number of acres still varies directly as the length, depending, for any given width, on that alone.

4. Now, when we have to calculate the right of way or clearing for a located line from 10 to 200 miles long, even this simple diagram may become a very considerable convenience, and can be constructed in a few minutes. On a sheet of ordinary cross-section paper the lines *O D*, Fig. 2, for the widths required, are first pencilled on with great care and accuracy, several points on each line being determined. Then, with the scales assumed above, each of the fine vertical lines represents some even 10 feet in length of centre line. When, as usually happens, that length is *not* some even 10 feet, say 1,437 feet, we *imagine* a line to be interpolated between those for 1,430 and 1,440 feet, and follow up that imaginary line till it cuts the line *O D*, Fig. 2, at a point. We have then only to read off from the horizontal lines the length by scale of that vertical; the heavy horizontal lines representing some even quarter of an acre, and the fine lines some even twentieth, or .05, of an acre. If the point above falls directly on one of those lines, we read off the value of that line at once. If otherwise, we first find the value of the horizontal line just below the point, and estimate by the eye the remainder of the distance. Thus, in Fig. 3, the area due to a strip 100 feet wide and 3,000 feet long is seen to be about 6.9 acres; and for a strip 2,600 feet long, 6 acres. If Fig. 3 were full scale, the area could be easily read within five-thousandths of an acre, and even nearer if the diagram be constructed on profile paper. The process is exactly analogous to reading off from a scale the length of a line on a drawing.

5. The construction of diagrams for *earthwork* computation is somewhat less simple than the above instance; since we then most frequently find that, instead of having only *two* variable quantities, one of which depends directly on the other, we have *three*, and that the value of the one sought depends upon the *other two;* as if, to continue our illustration, the width were liable to vary with each case as well as the length;—as indeed might be the case on some English railways, where the land is not taken of any particular width, but the fence set a certain distance outside of the slope-stakes.

6. This difficulty is met as follows: It has been already stated that, when we have but *two* variable quantities, the necessity of drawing special lines for particular values of each is avoided by previously drawing on lines for certain even values. When we have added a third variable quantity, this method is extended *to the inclined lines*, which (see Fig. 3) represent values of that quantity; that is to say, we draw in the inclined lines shown in Fig. 3 for certain even values of the third variable quantity, which will bring them so near together that the position of any intermediate line can be estimated by the eye without sensible error. For example, in the previous illustration, lines would be plotted for widths of 101, 102, 99, 98, 97 feet, etc.; and we thus obtain a diagram from which the area may be read off not only for a strip of any *length*, but of any *width* also. An inspection of Fig. 7, page 20, or any of the plates, will render this method clear.

It may seem that this additional interpolation introduces too great an element of uncertainty, but, after a few minutes' practice, the process becomes mechanical; and, in most of the diagrams, the only interpolation necessary is for a single line, half-way between those actually drawn. This may be tested on Fig. 7, page 21, by taking off the numerical examples given in Fig. 8 and paragraph 21, or by constructing any number of solids similar to Fig. 8.

---

7. The above explanation has been made without any reference to Analytical Geometry, and the diagrams might be constructed and used without any knowledge of it. Their construction, however, is really based on the elementary principles of that study, and is explained more concisely and clearly by using some of its technical terms. As a matter of convenience, it will be hereafter assumed that the following points are known to the reader.

8. In Analytical Geometry all points or lines are located with reference to two lines, usually at right angles to each other, termed the coördinate axes; their point of intersection, O, Fig. 3, being known as the origin of coördinates, or simply the origin, the horizontal line being known as the axis of x, or axis of abscissæ, and the vertical line as the axis of y, or axis of ordinates. Distances laid off along each are termed respectively abscissæ and ordinates, and points which have their abscissa and ordinate given are located by having those values laid off on the proper axis, and, through the points thus obtained on each axis, drawing lines parallel with the other axis; the intersection of these two lines locating the point.

Every equation consists of constants, or known quantities, and variables, or unknown quantities, and any equation of the first degree containing but two variables is the equation of a straight line; that is to say, can be graphically represented by a straight line which will pass through every point we should obtain by substituting special values of one of the variables in an equation, and deducing therefrom the value of the other. Every equation of the second or higher degree, on the contrary, can only be represented by a curve.

The quantity of which values are laid off on the horizontal axis is often termed simply the variable x; and that of which values are laid off on the vertical axis, the variable y; and the line, whether straight or curved, which represents an equation, is termed the *locus* of that equation.

9. Now, referring to the example on page 1, we see at once that, letting $l =$ the length of the strip, and $A =$ the corresponding area in acres, it can be expressed in the form of the equation—

$$A = \frac{100}{43560} l. \qquad (1),$$

which is of the first degree between two variables, A and l, and hence can be represented by a straight line. To construct this line, we substitute special values of l in equation (1), and, through points obtained by plotting the values then deduced from the equation, pass a right line, every point on which *satisfies* equation (1); and hence, if we have given the value of one of our variables, as l, we can determine the value of the other by measurement or inspection. When we have several different widths, we have of course a different equation, and consequently a different line, for each; but these lines will all pass through the origin, O, Fig. 3, since, when $l = 0$, the equation in all cases reduces to zero.

When, however, the *width* also is variable, equation (1) takes the form—

$$A = \frac{w}{43560} l. \qquad (2),$$

in which there are *three* variables, and hence it cannot be plotted as a line; but we are enabled to solve it graphically by treating w as a constant, and assigning successive values to it differing by a small amount, as one foot, for each of which values we obtain a different line from the equation; these lines being then so near together that the position of lines for intermediate values of w can be estimated by the eye.

10. It is sufficiently evident from the above that it may be stated, GENERALLY, that *any* equation between three variables can have a diagram thus constructed for it, by the aid of which it may be solved at once by inspection; and, if the equation is of the first degree, all the lines of the diagram will be straight. When an equation contains more than three variables, a diagram cannot be constructed for it directly.

11. All the equations required for the practical computation of earthwork are of the former description; that is to say, they involve, or by a slight change in the formula can be made to involve, but two variable quantities besides the quantity sought, and hence they can be solved at once, on diagrams constructed for that purpose, without further computation, for equidistant stations, than simple addition.

The preceding explanation, and a few minutes' practice with the numerical examples given for each diagram, will enable any one to take off quantities from any of the plates. They are all constructed and used in the same manner, and values of the quantity sought are always read off from the horizontal lines.

12. In the following chapter, the diagrams constructed to solve formulæ for the following purposes, are taken up in order:

First. The computation of end-area solidities, for "three-level" sections.

Second. The computation of "three-level" earthwork by the prismoidal formula.

Third. The computation of earthwork having "five-level," or other irregular sections of any form.*

Fourth. The computation of solidities by the formulæ given in Henck's "Field-book for Railroad Engineers."

Fifth. The determination of the correction due to curvature.

Sixth. The computation of preliminary estimates.

The mathematical formulæ are first determined, and the final equations, which the diagrams are constructed to solve, distinguished by a ☞. The special rule for use and applications of each diagram are then given. The manner of constructing diagrams is considered more minutely in Chapter IV.

13. It has been assumed in constructing the diagrams that the eye will divide the space between the fine horizontal lines into tenths. Some persons are deficient in this faculty, but it is readily acquired by practice. It is always best for two to work together in using a

* "Five-level" cross-sections are thus classified for convenience, though they cannot strictly be called irregular.

diagram ; one to call off from, and set down results in the note-book, and the other using the diagram. The rate of progress will commonly depend on the former, as end-area solidities can be taken off in ten or fifteen minutes for a whole day's cross-sectioning. A needle-point should be used to determine and hold the point of intersection, while reading off from the horizontal lines.

NOTE.—Wherever the term "station," or "full station," is used in the following pages, it is assumed to be 100 feet long.

The term "prismoid" has been—for convenience—frequently applied to solids bounded by warped surfaces. This, however, is of course not justified by the mathematical definition of that solid, which requires that all the bounding surfaces shall be planes.

CHAPTER II.

FORMULÆ AND RULES.

ART. I. — *Computation of End-Area Solidities, for "Three-Level" Sections.*

14. THE field-notes give the following data: The centre height, the distance out from the centre to the slope-stake on each side, and the height of each slope-stake above or below grade. These notes are here assumed, for convenience, to be recorded in this form:

$$\frac{17.8}{-7.2}, \quad -8.8, \quad \frac{22.3}{-10.2},$$

the upper figures denoting the distance out of the slope-stake, and the lower ones the cut or fill on the centre and slope-stakes respectively. The form of keeping the notes is, of course, immaterial.

15. The solidity by the Method of End Areas is calculated from these notes by the following rule: Compute the area of the section at each end of the solid; add them together and divide by two; multiply the quotient by the length of the solid, to give the solidity in cubic feet; and divide by 27, to give the solidity in cubic yards.

To compute the sectional area, the section is supposed to be divided into four triangles, as indicated by the dotted lines in Fig 4, and its area is then computed

FIG. 4.

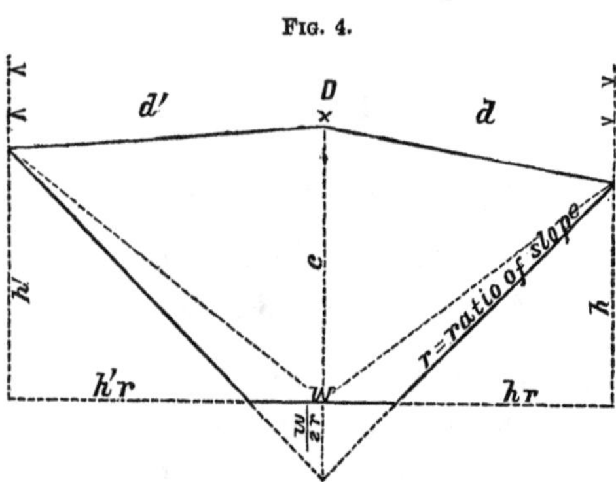

as follows: Take the sum of the two distances out, $(d + d')$, or D; multiply it by the centre height, c, and reserve the product. Also, take the sum of the two side heights, $(h + h')$, and multiply it by half the width of the road-bed, w. The sum of these two products, divided by 2, is the area.

Diagram of Cross-Sections.

16. Expressed by formula, the above rule is—

$$A = \frac{(d + d')\,c + (h + h')\,\frac{w}{2}}{2}. \qquad (3).$$

This equation contains four variables, A, c, $(d + d')$, and $(h + h')$; but it will be seen at once that d and h are what are termed *functions* of each other; that is to say, they depend on each other for their value, and if the value of the one is changed, that of the other is also changed correspondingly. In all cases—

$$d = hr + \frac{w}{2}; \text{ and } h = \frac{d - \frac{w}{2}}{r}.$$

Substituting this value of h in equation (3), we have—

$$A = \frac{(d + d')c}{2} + \frac{\left(\frac{d - \frac{w}{2}}{r} + \frac{d' - \frac{w}{2}}{r}\right)w}{4},$$

$$= \frac{(d + d')c}{2} + \left(\frac{d + d' - w}{r}\right)\frac{w}{4},$$

$$= (d + d')\frac{c}{2} + (d + d')\frac{w}{4r} - \frac{w^2}{4r},$$

or, letting $d + d' = D$,

$$A = \left(\frac{c}{2} + \frac{w}{4r}\right)D - \frac{w^2}{4r}. \qquad (4).$$

We can obtain this formula in another way, which is perhaps more obvious, thus: Extend the side-slopes, and the centre line, c, Fig. 4, till they intersect in a point. The length of the prolongation of the centre line is $\frac{w}{2r}$, and the area of the added triangle* below the road-bed is $\frac{w^2}{4r}$. Obviously, the area of the section

* This added triangle is often termed the "grade-triangle," and the solid due to it the "grade-prism." They will be so termed throughout this volume.

is equal to the entire length of the centre line thus prolonged, multiplied by $\frac{D}{2}$, less the area of the added triangle. Putting this in the form of an equation, we have, as before,

$$A = \left(\frac{c}{2} + \frac{w}{4r}\right) D - \frac{w^2}{4r}. \qquad (4).$$

We have now to combine with equation (4) the remaining steps in the rule for computation, in order to obtain the volume at once.

Letting $S =$ the end-area solidity of a solid of the usual unit of length, 100 feet,

$$S = \frac{A + A'}{2} \times \frac{100}{27},$$

$$= (A + A')\frac{100}{54}.$$

By constructing the diagram to give areas already multiplied by the factor $\frac{100}{54}$, we shall get the solidity for a full station by the simple addition of the diagram quantities for any two end sections. Multiplying equation (4) by this factor, and letting $A \times \frac{100}{54} = s$, we have—

☞ $$s = \frac{50}{54}\left(c + \frac{w}{2r}\right) D - \frac{25w^2}{54r}; \qquad (5).$$

s being equal to the solidity of a prism of the given section fifty feet long.

FIG. 5. OUTLINE OF DIAGRAM OF CROSS-SECTIONS, ROAD-BED, 18; 1½ TO 1.
Showing the Method of Condensation on the Plate.

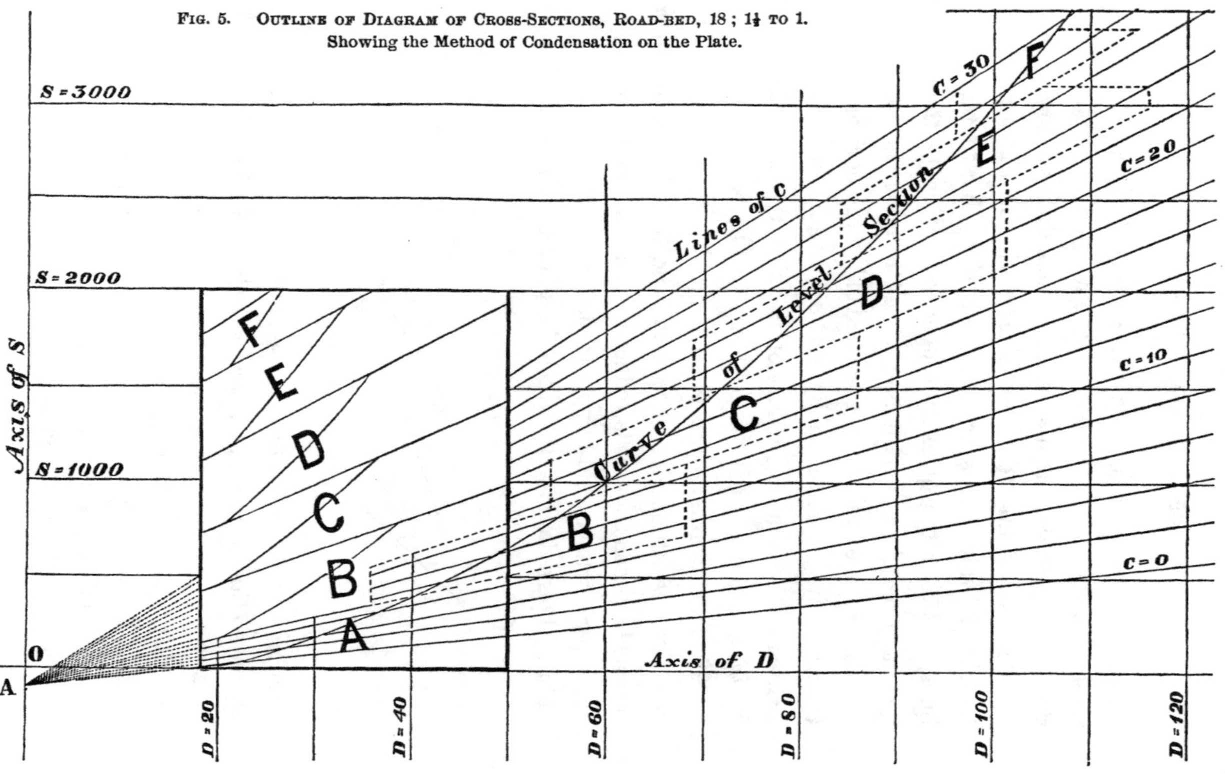

17. In this equation (5), there are but three variables: c, the centre height; D, the width from slope-stake to slope-stake; and s, the corresponding solidity for a length of fifty feet. In constructing a diagram to solve it, values of s are represented by horizontal lines, as being the quantity sought, according to the general rule given in Chapter I.; D is taken as the variable x; i. e., values of it are laid off on the horizontal axis, and represented by *vertical* lines; and c is treated as the constant, values of it being represented by *inclined* lines. All this is shown in Fig. 5.

18. The diagram presents one peculiarity in construction, which is also shown in Fig. 5. To avoid unwieldy dimensions, it is reduced in size by omitting those parts which would apply to such unusual sections as are shown in Fig. 6, which in practice never, or

FIG. 6.

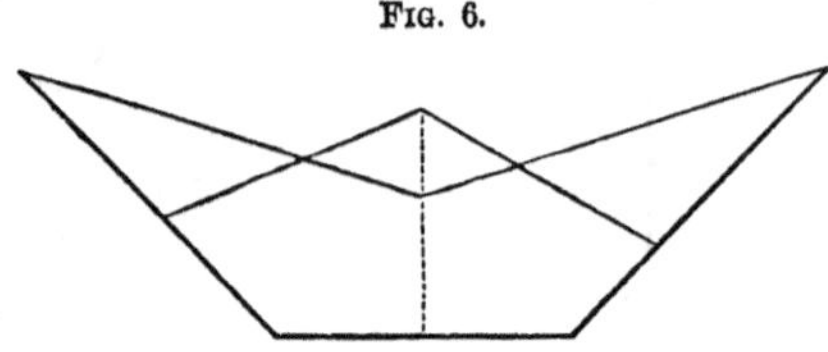

very rarely, occur. The values of c and D, in any given section, rarely differ *very widely* from those due to a section level on top, and hence only that part of the diagram which is inclosed in the dotted trapezoids in Fig. 5 is actually constructed. These separate trapezoids combine very conveniently on a single sheet of cross-section paper, as indicated in the figure, but the parts A, B, C, D, E, F, are entirely independent, and are separated from each other on the full-scale diagrams by a heavy black line. It is plain that the same horizontal or vertical line may have an entirely different value in a different subdivision.

19. We will now consider the case of SECTIONS LEVEL ON TOP, the solidity of which is sometimes required. In such sections—

$$D = 2cr + w.$$

Substituting this value in equation (5) we have—

$$s = \frac{50}{54}\left(c + \frac{w}{2r}\right)(2cr + w) - \frac{25w^2}{54r},$$

☞ $$= \frac{100}{54}(cw + c^2r). \qquad (6).$$

20. This is an equation of the second degree between two variables, and its locus is a parabola or single curved line. This curve (see Figs. 5 and 7) intersects each line of c at its intersection with the corresponding value of D due to a level section, or $D = 2cr + w$, and it affords the means of determining the solidity due to level sections when either the width on top or the centre height is given. It also furnishes a rapid method of reducing any section to an equivalent level centre height. It is drawn on all the diagrams, though more as a matter of convenience than necessity; and, in taking off the solidity of "three-level" sections, is disregarded.

21. Rule for Use of Diagram.—First add together the distances out of the two slope-stakes, to obtain the value of *D*. Then *follow up the vertical line representing the given value of D to its intersection with the* INCLINED *line representing the given centre height, or value of c. Hold the point of intersection, and read off the solidity for a half-station from the horizontal lines.*

The sum of the quantities for any two sections is the end-area solidity between them for a full station. For fractional or *plus* stations, multiply that sum by the proper decimal part of 100 feet. A tabular example of the form of keeping notes is given in Chapter IV., page 110.

Example.—To compute the solid shown in Fig. 8. For the first section, take on Plate III., or on Fig. 7, which is an extract from Plate III., the vertical line marked 45—the value of *D* for the first section—and follow it up to the inclined line representing the given centre height, 8.4. A heavy line marked 8 is first met with, the next line above is 8.2, and the *next* line is 8.4. Place the needle-point on the intersection of these two lines, and read off the solidity from the horizontal lines.

In this case the point of intersection lies exactly on the line 500; which is the solidity required. Had the centre height been 8.6, the solidity would have been 508; had the centre height been 8.8, the solidity would

FIG. 7.

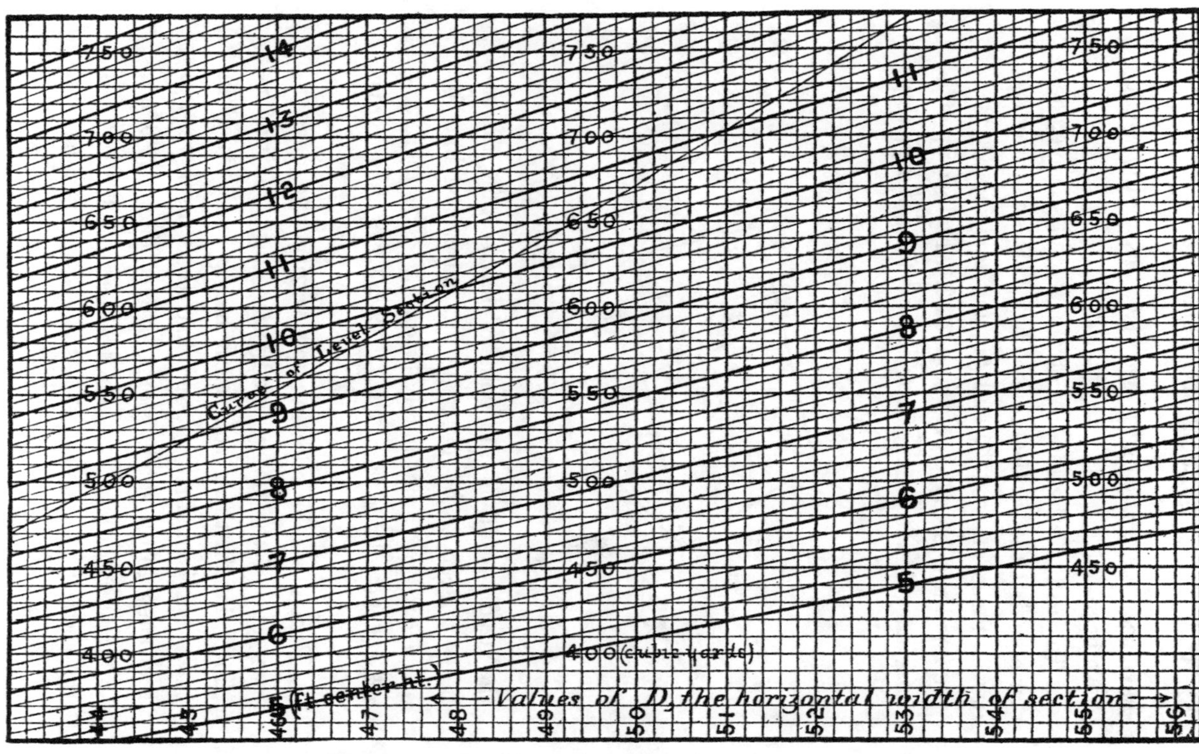

EXTRACT FROM DIAGRAM OF CROSS-SECTIONS; ROAD-BED, 18, 1½ TO 1.

have been 517; and had it been 9.0, the solidity would have been 525.

To read off the solidity of the second section; determine first the vertical line 55.5. This is midway between those marked 55 and 56, and midway between the fine lines for $D = 55.4$ and 55.6. Follow up this imaginary line to the inclined line $c = 9.8$; and placing

Fig. 8.

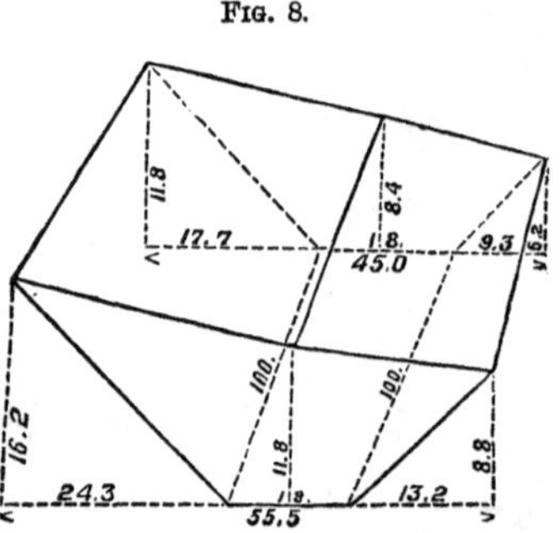

the needle-point on the point of intersection, the solidity is seen to be 712; the fine horizontal line for $s = 710$ being immediately below it. The total volume of the solid is then $500 + 712 = 1{,}212$ cubic yards.

Remember, in following up a vertical line, never to *cross* one of the heavy subdivision lines. If the inclined line for the given centre-height is not found in the subdivision first entered, find the proper vertical line anew in the subdivision above or below it. Thus, in taking off from Plate III. the first section above: If the vertical line marked 45 *at the bottom* of the diagram be taken, and followed up—without regard to the heavy black line—to the inclined line 8.4, the quantity obtained will be 740, instead of 500. This is so obvious as only to require mention, and the process soon becomes mechanical; so that the needle-point can be placed almost at once on the required intersection. It is convenient to have the values of D in the note-book already computed; as in the tabular form on page 110.

22. SECTIONS OF OVER THIRTY FEET CENTRE-HEIGHT may be taken off as follows:

Subtract from the centre-height the altitude of the grade-triangle, $\frac{w}{2r}$, Fig. 4, as given in the table below. Then, divide both the centre-height and width on top by 2, and take off a quantity for the new notes from the diagram. Multiply it by 4, and add to the product the constant quantity given below for the road-bed and plate in use, viz.:

For a Road-bed of	Plate.	Subtract from the Centre-height.	And add to *four times the quantity taken off.*
14; $1\frac{1}{2}$ to 1	I.	4.67	181.5
15; " " "	II.	5.	208.3
18; " " "	III.	6.	300.
20; " " "	IV.	6.67	370.4
18; $1\frac{1}{4}$ to 1	V.	7.2	360.
18; 1 to 1	VI.	9.	450.

The quantities to be added—in the last column of the table above—are equal to the solidity of the grade-prism * of the given road-bed for a length of 150 feet.

The principle of this process is very simple, and may be indicated as follows: If the triangle below the road-bed were *included* in the section, dividing each dimension by 2 would reduce the area to one-fourth of the truth. Also, the centre-height, as obtained above, will be found to be equal to half the centre-height, if it included that triangle, *less the altitude* of the triangle. It follows, that the quantity taken off from the diagram, added to that due to the grade-triangle, will be one-fourth of the solidity due to the original section with the grade-triangle included in it; and hence, *four* times the quantity taken off plus *three* times the solidity of the grade-prism (equal to its volume for a length of 150 feet) gives the correct solidity due to the original section. This process doubles the range of the diagram with but little extra labor.

EXAMPLE.—A section in which $c = 48.4$, $D = 172.0$; — road-bed 18, $1\frac{1}{2}$ to 1. From the table above, the altitude of the grade-prism is 6.0. Then $48.4 - 6.0 = 42.4$; $\div 2 = 21.2$. Entering the diagram with $D = \frac{172.0}{2} = 86.0$, and $c = 21.2$, we obtain 2066; $\times 4 = 8264$; $+ 300$ (from the table above) $= 8564$.

The error of observation is multiplied by 4 by this process, and will sometimes cause an error of from one

* See Fig. 4, and foot-note to page 15, for a definition of this term.

to three yards—but this is unimportant on such large sections.

23. FOR A SECTION OF UNUSUAL SHAPE, WHICH CANNOT BE DIRECTLY TAKEN FROM THE DIAGRAM, owing, not to the size of the section, but to the curtailment of the diagram indicated in Fig. 5; the preceding method may be used, but the following is slightly more accurate and convenient:

Add to, or subtract from, the centre-height, as the case may require, 5.4 feet; the section can then always be taken off. From the solidity obtained, subtract, or add, as the case may be, $5D$, or five times the width of the section on top.

The principle of this process is obvious. By the subtraction of 5.4 from the centre-height, the area of the section is diminished by $5.4 \times \frac{D}{2}$, and consequently the solidity by $\frac{100}{54} \times 5.4 \times \frac{D}{2}$, or $5D$; which amount has then to be added to obtain the true solidity of the section. In this way also the diagram can be extended to several feet greater centre-heights than are shown on it. All such sections, however, can be computed also from the Diagram of Triangular Prisms, if preferred, as stated in paragraph 53.

EXAMPLE.—Take the following section, road-bed 18, 1½ to 1:

$$\frac{16.8}{+5.2} \qquad +7.4 \qquad \frac{15.6}{+4.4}.$$

In this section $D = 32.4$. Entering Plate III. with $D = 32.4$, $c = 7.4$, the section cannot be taken off. Subtracting, then, 5.4 from the centre-height, we have $c = 2.0$. Entering the diagram with this *new* centre-height, we obtain 140. To this is to be added $5D$, $= \frac{324}{2}$, $= 162$; giving 302 as the solidity due to the section.

24. Sections can be reduced to an equivalent level centre-height, by means of the Curve of Level Section, as follows: After having determined the point of intersection for the values of D and c in any given case, and the horizontal line on which it lies; follow along that horizontal line till it intersects the Curve of Level Section. The inclined line of c, interpolated if necessary, which passes through this point of intersection, gives the centre-height required.

The principle of this is obvious. The two sections must necessarily have the same area, since the points of intersection found for them are on the same horizontal line.

Example.—Take the first section of Fig. 8. After determining, on Fig. 7 or Plate III., that the point of intersection, as obtained in paragraph 21, lies on the line 500, follow along that line till it cuts the Curve of Level Section. The inclined line passing through the point of intersection is read off as 8.7; which is the equivalent level centre-height. In the second section, the equivalent level centre-height is read off in the same manner as 11.1.

25. The Curve of Level Section furnishes a check against gross errors in the notes, as follows: If, in a given section, the value of D is *greater* than that due to a section level on top, as in the larger section shown in Fig. 9, the line for that value of D on the diagram

Fig. 9.

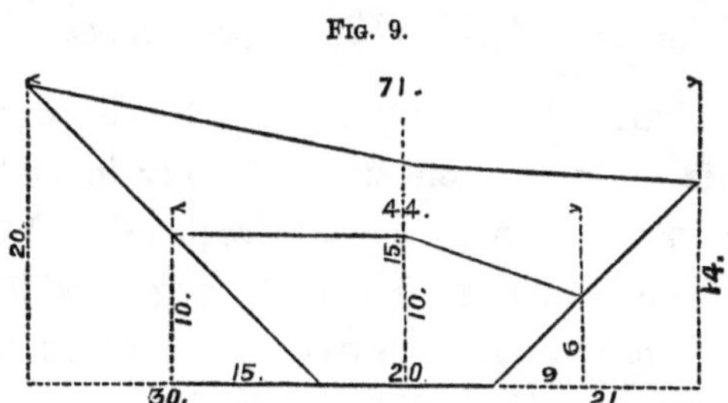

will evidently intersect the line for the given value of c *beyond*, or to the right of, the Curve of Level Section; but if it be *smaller*, as in the smaller section in the same figure, then it will intersect to the left of the curve; and, on nearly level ground, all the points of intersection found will lie close to the curve. Then, knowing the general character of the surface, we have a guard against gross errors, either in the field-work or in the office. In a proximately level country, an error of a foot will attract attention.

26. The reader will sometimes be in doubt, when taking off quantities, as to which *is* the nearest yard; but, to illustrate the small importance of the decision, suppose the width on top of the lower section of Fig. 8 had been recorded as 55.4, instead of 55.5. The solidity is then taken off 710, a difference of two cubic yards. Such a variation in the notes is the merest accident—not to speak of subsequent variations in the execution of the work.

Diagram of Side-Hill Cross-Sections.

27. Sections technically termed "side-hill," one side of which is in excavation and the other in embankment, are met with more or less frequently. Such a section is shown in Fig. 10, the lettering of which requires no explanation. They can be taken off by inspection from a diagram, but the parts A, B, and C must be given separately, since such sections consist of two parts, A and BC, which are in reality independent; having the common centre-height, c, but different road-beds and slopes. The side $B\ C$ also consists of two parts, one in excavation and one in embankment, which must be given separately. After determining the solidities due to A, B, and C, A and B are combined.

FIG. 10.

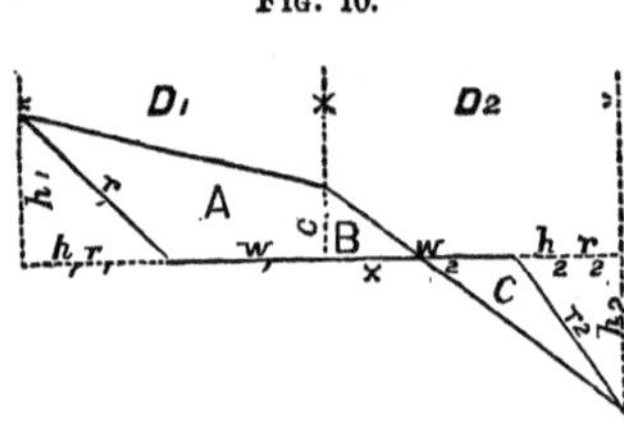

28. We have first to obtain the formulæ of area, which will be so expressed that the diagram may be constructed in terms of c and D, and so be similar, in use and general design, to that for thorough-cut sections.

The equation of the trapezium A is very similar to that of a thorough-cut section. With the lettering of Fig. 10—

$$A = \frac{c}{2} D_1 + \frac{h_1}{2} w_1,$$

but $h_1 = \frac{D_1 - w_1}{r_1}$, and substituting this value, we have

$$A = \left(\frac{c}{2} + \frac{w_1}{2r_1}\right) D_1 - \frac{w_1^2}{2r_1}.$$

Multiplying this equation by $\frac{100}{54}$, as in the diagram for thorough-cut sections, we have—

☞ $$sA = \frac{50}{54}\left(c + \frac{w_1}{r_1}\right) D_1 - \frac{50 w_1^2}{54 r_1}. \qquad (7).$$

For the triangle B, if we let x = its base, we have $B = \frac{cx}{2}$. To determine the length of this base, x, in terms of our given dimensions, we have, by a proportion evident from Fig. 10—

$$x = \frac{cD_2}{c + h_2} = \frac{c(w_2 + h_2 r_2)}{c + h_2}.$$

Substituting this value of x, and multiplying by $\frac{100}{54}$ as before, we have—

☞ $$sB = \frac{100}{54} \frac{c^2 (w_2 + h_2 r_2)}{2(c + h_2)}. \qquad (8).$$

For the triangle C we have—

$$C = \frac{(w_2 - x)\,h^2}{2};$$

or, substituting the value of x above—

$$C = \left[w_2 - \frac{c\,(w_2 + h_2 r_2)}{c + h_2}\right]\frac{h_2}{2}.$$

Reducing and multiplying by $\frac{100}{54}$, we have—

☞ $$sC = \frac{100}{54}\,\frac{h_2{}^2(w_2 - cr_2)}{2\,(c + h_2)}. \qquad (9).$$

29. The above equations, (8) and (9), for the triangles B and C, are expressed in terms of c and h_2. They could now be expressed in terms of c and D_2, in terms of which the diagrams are constructed, by simple substitution, since h_2 is a function of D_2; but the above forms are simpler of solution, and the substitution is more easily made by solving the equations for values of h_2 corresponding to given values of D_2. The equations being of the second degree, the lines of c, in the diagrams of B and C, will be curved, but lines for even feet only are drawn on, since such sections are always small. The diagram of A, based on equation (7), consists of right lines, as in the thorough-cut diagram, to which it is very similar.

Fig. 11 is an outline of a diagram which would be sufficient to solve all side-hill sections for road-beds of 18 and 14 feet respectively, and side-slopes of 1½ to 1.

Fig. 11.

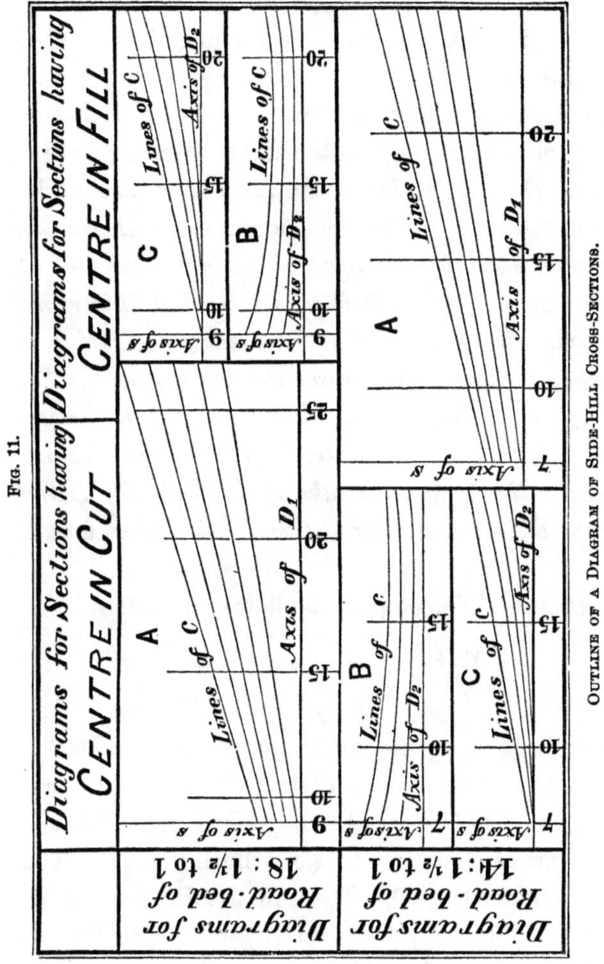

Outline of a Diagram of Side-Hill Cross-Sections.

The only peculiarity in construction is the manner of arranging the diagrams for *A*, *B*, and *C*. It will be seen that those for each *road-bed* are in a horizontal column, but those applying to *any given section* are in the vertical column headed "Centre in Cut" or "Centre in Fill."

30. RULE FOR USE OF DIAGRAM.—This is the same as that for the thorough-cut diagram, except that the value of *D on each side* of the centre-line is used, instead of the entire width from side to side. *Beginning always on the "thorough-cut" side of the section*—that is, with *A*, Fig. 10—*follow up the vertical line representing the distance out of the slope-stake on that side, to the inclined line representing the centre-height. The solidity due to A is then read off from the horizontal lines. Then, with the centre-height and the distance out of the other slope-stake, take off the solidity of B and C from their respective diagrams, and add A and B together.*

EXAMPLE.—To compute the following section :

$$+\frac{18.4}{6.3} \qquad +1.6 \qquad -\frac{12.4}{3.6},$$

the road-beds being 14 and 18 feet respectively, which are the two at the top of Plate VII. Under column "Centre in Cut," enter the diagram of *A*, and follow up the vertical line 18.4 to the inclined line 1.6. The solidity obtained is 80. Then, in the Diagram of *B*, just below, follow up the vertical line 12.4 to the inclined line 1.6. The solidity obtained is 6, which is added to the 80; and, with the same notes of $D = 12.4$, $c = 1.6$, the solidity due to *C* is read off from its diagram.

On the sheet of side-hill diagrams a number of different road-beds are given; but only two of these are required on a given piece of work, as is evident. The others must be disregarded, and the two in use imagined to be brought together, if not already so, in the manner indicated by Fig. 11.

31. THE GRADE-POINT IS SOMETIMES TAKEN, AND ITS DISTANCE FROM THE CENTRE RECORDED, in side-hill sections. The section then becomes "four-level," and the solidity for the triangles B and C must be taken from the Diagram of Triangular Prisms; that of A only being taken from the Diagram of Side-Hill Cross-Sections. The practice offers no special advantage unless the section is irregular.

ART. II.—*Computation of "Three-Level" Sections by the Prismoidal Formula.*

32. The method most commonly used for the accurate computation of earthwork is by means of the PRISMOIDAL FORMULA; which, stated in words, is as follows:

Add together the areas of the two end-sections, and four times the area of a section parallel to and midway between the two end-sections, and multiply by one-sixth the length. The product is the solidity.

Expressed algebraically, letting A and $A' =$ the

area of the end-sections, $M =$ the area of the mid-section, and $S =$ the solidity, we have—

$$S = (A + A' + 4M)\frac{l}{6}. \qquad (10).$$

This formula applies to all solids having parallel ends, and bounded by plane surfaces.

33. The surface of an earthwork solid, however, is rarely a perfect plane, but the two end-sections may be said to *always* differ, more or less, in ground-slope, as indicated in Figs. 9 and 12. In such cases it is assumed, in

Fig. 12.

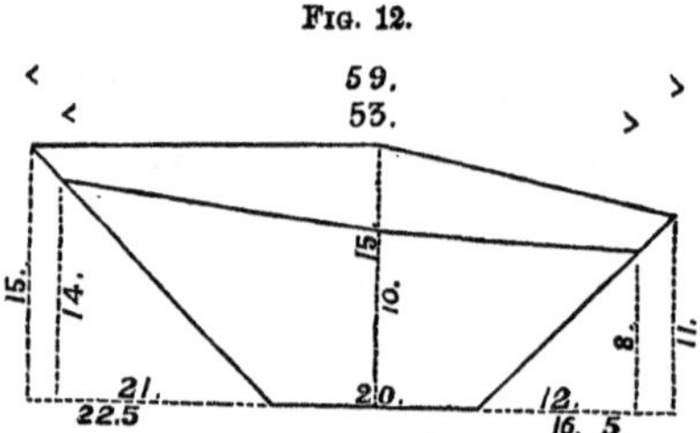

applying the prismoidal formula, that the edges of the solid are right lines, and that a section of the solid at any point, parallel with the end-sections, is *also* a three-level section, and bounded by right lines. There is but one other assumption as to the surface which can properly be made, namely, that which forms the basis of Henck's system of computation,* that the two upper

* "Field-book for Railroad Engineers," by John B. Henck, A. M., pp. 99, 100. The curious in such matters will find an elaborate discussion of the probabilities as to surface in Prof. Gillespie's "Roads and Railroads," rev. ed., p. 356. See also "New Theorems, Tables, and Diagrams for the Computation of Earthwork," by John Warner, A. M., p. 16, *et al.*

surfaces of the prismoid, on either side of the centre-line, may each be regarded as made up of two triangular plane surfaces. This assumption is separately considered in Art. IV. of this chapter.

34. Under the former assumption, each of the two top surfaces of the solid, on each side of the centre-line, becomes a HYPERBOLIC PARABOLOID; defined in geometry as *the surface generated by a right line*—here the surface lines of the cross-section—*moving parallel with itself along two other right lines not in the same plane*—here the longitudinal edges of the solid; and the late Prof. Gillespie has demonstrated that the prismoidal formula applies to solids bounded by such surfaces.* The mid-section, by this method, will be given by taking the half-sum of all the corresponding dimensions of the end-sections, as is sufficiently evident; and it is taken by all writers as the standard for accuracy in computing by the prismoidal formula.

35. This method of computation is rarely used, owing to the labor required in determining the mid-section, and, instead of it, that known as the "METHOD OF EQUIVALENT LEVEL SECTIONS" is more commonly employed; which makes no assumption whatever with regard to the surface, but determines the mid-section as follows: The centre-height of a *level* section, equivalent in area to each of the two end-sections, is first determined; and the average of these two heights is assumed

* "Roads and Railroads," p. 367.

to be the equivalent level centre-height of the mid-section.

The area of the mid-section, as thus determined, is always smaller than by the first method, except when the surfaces of the end-sections are "similar," so that the solid is bounded by plane surfaces.* The method of "Roots and Squares," so called, and those of Macneill, Baker, Trautwine, Lyon, Rice, and almost all other compilers of tables, are merely this method in more or less modified forms.

36. For each of these methods, a modification can be made—which is much more convenient for computation by diagrams—by which we are enabled to determine *the amount of error involved in computing a given solid by "end-areas,"* instead of directly computing by the prismoidal formula. This gives the same result in the end, but simplifies the process, and enables end-area solidities to be first obtained, and the error therein involved to be subsequently determined at any time, for an accurate final estimate. These formulæ, indeed, are more convenient for numerical computation also.

Diagram of Prismoidal Correction.

37. Under the first method given above, that of DIRECTLY DETERMINING THE DIMENSIONS OF THE MID-SECTION, the formula of correction is obtained as follows:

* For a demonstration of this fact, see Appendix A.

In any prismoid, letting A and $A' =$ the areas of the two end-sections, we have, by equation (4),

$$A = \frac{c}{2} D + \frac{w}{4r} D - \frac{w^2}{4r}; \qquad (11).$$

$$A' = \frac{c'}{2} D' + \frac{w}{4r} D' - \frac{w^2}{4r}. \qquad (12).$$

First determining the "end-area" solidity, S_E, by adding equations (11) and (12) together, dividing by 2, and multiplying by l, we obtain—

$$S_E = \left(\frac{c}{2} D + \frac{c'}{2} D' + \frac{w}{4r}(D + D') - \frac{2w^2}{4r}\right)\frac{l}{2},$$

$$= \left(\frac{3c}{2} D + \frac{3c'}{2} D' + \frac{3w}{4r}(D + D') - \frac{6w^2}{4r}\right)\frac{l}{6}. \quad (13).$$

We have now to obtain the true solidity, S, to do which we must first obtain the equation of area for the mid-section, by taking the half-sum of the corresponding dimensions of the two end-sections in equations (11) and (12). We thus obtain—

$$M = \frac{c + c'}{4} \frac{D + D'}{2} + \frac{w}{4r} \frac{D + D'}{2} - \frac{w^2}{4r},$$

and hence—

$$4M = (c + c') \frac{D + D'}{2} + \frac{2w}{4r}(D + D') - \frac{4w^2}{4r}.$$

Substituting the above values of A, A', and $4M$ in the prismoidal formula, equation (10), we have—

$$S=\left(\frac{c}{2}D+\frac{c'}{2}D'+(c+c')\frac{D+D'}{2}+\frac{3w}{4r}(D+D')\right.$$
$$\left.-\frac{6w^2}{4r}\right)\frac{l}{6}. \qquad (14).$$

Subtracting equation (13) from equation (14), terms containing w and r disappear, and we have—

$$S_E-S=\frac{cD+c'D'-c'D-cD'}{2}\frac{l}{6};$$

OR, letting $S_E-S=C=$ the amount of error in the solidity by the "Method of End-Areas"—

$$C=(c-c')(D-D')\frac{l}{12}. \qquad (15).$$

To adapt this formula to the construction of a diagram, 100 is substituted for l, and the last member of the equation divided by 27, in order to obtain the correction at once in cubic yards for a full station; giving to the equation the form—

$$C=\frac{c-c'}{3.24}(D-D'). \qquad (16).$$

38. This formula is independent of road-bed and slope, and applies generally to all three-level railway prismoids. It is the basis of the Diagram of Prismoidal Correction, the general arrangement and construction of which is shown in Fig. 13, and is similar to that of the Diagram of Cross-Sections and Triangular Prisms, in the assignment of the variables.

FIG. 13.

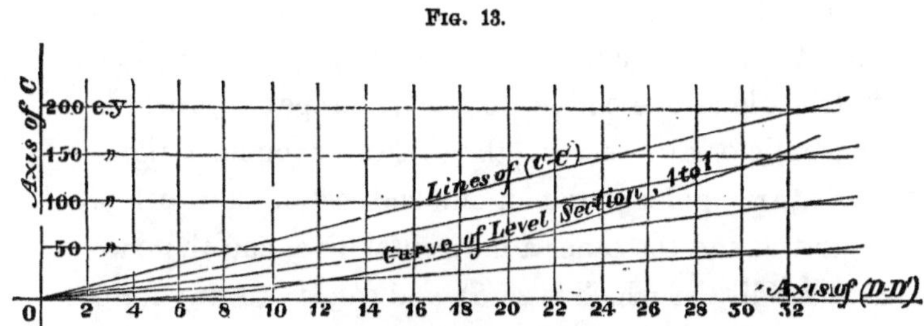

OUTLINE OF THE DIAGRAM OF PRISMOIDAL CORRECTION.

Equation (15) above applies also to *all* prismoidal solids of triangular section, whether bounded by plane surfaces or by hyperbolic paraboloids.* If it be multiplied by two, so as to give equation (15) the following form:

$$C = (c - c')(D - D')\frac{l}{6},$$

it applies to all solids of which the end-sections are rectangles or parallelograms, and—by averaging the parallel sides — to those which have trapezoids as one or both end-sections, provided the mid-section be also a trapezoid. And since a triangle may be regarded as a trape-

* A prismoid, of which the end and middle sections are dissimilar triangles (those in Fig. 16 for example), must necessarily have one or more warped surfaces—which are hyperbolic paraboloids: If the triangles are similar the solid becomes simply the frustum of a pyramid. So also a surface on *any* solid, every cross-section of which is bounded by right lines, must be a hyperbolic paraboloid, if *it* also be bounded by right lines which are not in the same plane. The prismoidal formula applies indifferently to such surfaces and planes.

zoid, and a line as a rectangle, of which one dimension is zero, the formula applies to all solids which have a line or triangle as one end-section, and a trapezoid, triangle or parallelogram as the other. For example, a right-angled triangular pyramid, one end-section of which is a vertical and the other a horizontal line, and the end-area solidity of which is zero; and so also to any form of irregular wedge. Some of these applications we shall have occasion to refer to.

39. Under the Method of Equivalent Level Sections, a similar formula of correction is obtained as follows:

Let $ABCD$ and $A'B'CD$, Fig. 14, be the two

Fig. 14.

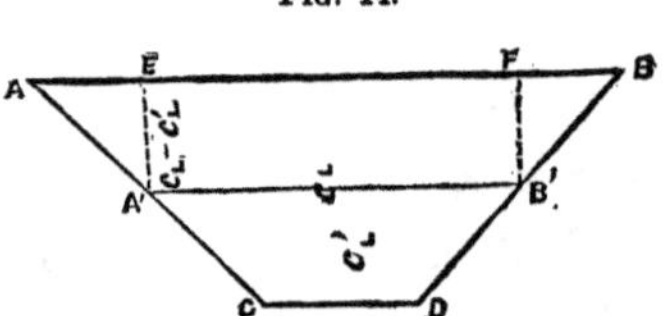

end-sections, reduced to level sections of equivalent area, of any prismoid. They are shown in the figure projected upon each other. Now, we can divide this solid into a prism, $A'B'CD$, a rectangular wedge, $A'B'EF$, and two triangular pyramids, $AA'E$ and $BB'F$. The solidity of the prism and the wedge are given correctly by averaging the end-areas, and all the error in so doing lies in the two triangular pyramids, of which the bases are $AA'E$ and $BB'F$. The area of each of these bases is $\frac{(c_L - c'_L)^2 r}{2}$, and the end-area solidity of both is, therefore,

$$S_E = (c_L - c'_L)^2 r \frac{l}{2}.$$

The *true* solidity of a pyramid, however, is given by the base multiplied by *one-third* the altitude, or—

$$S = (c_L - c'_L)^2 r \frac{l}{3}.$$

Subtracting this equation from the one above, and letting $S_E - S = C_L$, we have—

$$C_L = (c_L - c'_L)^2 r \frac{l}{6}. \qquad (17).$$

Substituting 100 for l, and dividing by 27, as before, we have—

☞ $$C_L = \frac{2r}{3.24}(c_L - c'_L)^2. \qquad (18).$$

40. This is an equation of the second degree between *two* variables, if r be constant, and its locus is a parabola. If this parabola be constructed on the Diagram of Prismoidal Correction, it cuts each of the inclined lines for values of $(c - c')$ at the intersection of that line with the corresponding vertical $(D - D') = 2r(c - c')$, as will be evident on comparing equations (16) and (18), and corrections by this method also may then be taken off from the same diagram. The locus of this equation is drawn in on Plate IX. for 1 to 1 and 1½ to 1 slopes, and is also shown on Fig. 13.

41. THE "METHOD OF CENTRE-HEIGHTS," so called, is sometimes used to determine a similar correction, under which it is assumed that the two end-sections may, *for the purpose only of obtaining this correction*, be regarded as level sections of their actual centre-height, and the correction is then determined by equation (18), as in the Method of Equivalent Level Sections. The locus of that equation on the Diagram of Prismoidal Correction is of course sufficient for computation by this method also. As equation (18), however, contains but two variables, it is readily tabulated in small compass, and such a table is added in Appendix B, to simplify somewhat the process of determining corrections by this method.

42. As compared with the Method of Level Sections, the Method of Centre-Heights is commonly less approximate for any one given solid; but the former, on the other hand, gives too small a solidity *in all cases*, when the solid is not bounded by plane surfaces, and, owing to the constant accumulation of these comparatively small errors, it gives in the end a not much closer approximation *in deficiency* than is obtained by the Method of Centre-Heights in excess. Hence the latter is preferable, even if it were not also the more rapid and simple.

The error, by either method, is of small intrinsic importance, and it is only a question of comparative expediency; but, for the benefit of those interested in exact computation, a more extended discussion of this question is given in Appendix A.

43. Rule for Use of Diagram.—*Take the vertical line representing the difference between the* WIDTH ON TOP *of the two end-sections, and follow it up to the inclined line representing the difference between the centre-heights. The correction for a full station is then read off from the horizontal lines.*

For fractional or *plus* stations, multiply the correction obtained by the proper decimal part of 100 feet. It is convenient to have the required differences already recorded, as shown in the tabular example of office notes, Chapter IV., paragraphs 112 and 115.

Example.—The sections shown in Figs. 9 and 12. In the first case, $D - D' = 27.0$, $c - c' = 5.0$. Follow up the vertical line 27 to the inclined line 5, and hold the point of intersection with a needle-point. It is found to lie at about $\frac{2}{5}$ of the distance from the horizontal line 40 to the line 45; giving 42 cubic yards as the correction.

In Fig. 12 we have $D - D' = 6.0$, $c - c' = 5.0$; and the corresponding correction, 9 + cubic yards. As the scale of the diagram is double that of the Diagrams of Cross-Sections, quantities may be read off to the nearest half-yard if desired, and the last quantity would then be taken off, 9.5.

44. To determine Corrections by the Method of Level Sections, each section is first reduced to an equivalent level section on the Diagram of Cross-Sections. Then, for a given solid, *follow along the curve for the given side-slope on the Diagram of Prismoidal Correction, till it intersects the inclined line representing the difference of centre-heights. Hold the point of intersection with a needle-point, and read off the correction from the horizontal lines.*

45. By the Method of Centre-Heights the process is the same as above, except that we simply take the difference of the *actual* centre-heights.

Example.—The sections shown in Figs. 9 and 12, as before. Reducing those in Fig. 9 to equivalent level centre-heights, by the Diagram of Cross-Sections (road-bed 20, 1½ to 1), we obtain 15.95 and 8.95. Following the curve of $r = 1\frac{1}{2}$ to 1 to the inclined line marked 7, we obtain 45.5. In Fig. 12, the equivalent level centre-heights are taken off as 14.0 and 10.5, and the correction obtained is 11.5.

By the Method of Centre-Heights, the correction obtained, in both instances, is 23—the difference of centre-heights, 5 feet, being the same.

These corrections may also be taken, somewhat more easily, from the table in Appendix B; as is there explained.

46. WHEN THE VALUE OF $(D - D')$ IS LESS THAN 4 FEET, multiply it by 10, and take one-tenth of the solidity obtained. The part of the diagram near the origin was omitted to avoid confusion of lines, but corrections for such values are usually unimportant.

WHEN THE VALUE OF $(D - D')$ EXCEEDS 44 FEET, divide it by 2, and double the correction obtained.

47. THE CORRECTION OBTAINED HAS SOMETIMES TO BE ADDED to the end-area solidity. This occurs when $(D - D')$ in equations (15) and (16) becomes a *minus* quantity; i. e., when the deepest cut is at the narrowest end of the solid. The two sections shown in Fig. 6 are an exaggerated instance. This, however, is a point of more curiosity than importance, as such corrections are always small and rarely met with. Similarly, when *either* $D = D'$ OR $c = c'$, $C = 0$; or, in other words, end-area solidities are then correct, however great the difference in the other two dimensions; yet there is always a subtractive correction by the Method of Level Sections, when the areas are not identical.

ART. III.—*Computation of Earthwork of Irregular Section.*

48. From the nature of irregular earthwork, no general formula can be devised which will enable complete solidities to be taken off from a diagram at a single operation; but by means of a very simple diagram the area multiplied by $\frac{100}{54}$ of any triangle or trapezoid—i. e., the solidity of prisms of such sections 50 feet long—may be read off at once, which reduces the computation of end-area solidities, for any solid whatever, to simple addition.

Diagram of Triangular Prisms.

49. In any triangle, letting $D =$ the base, $c =$ the altitude, and $A =$ the area:

$$A = \frac{cD}{2}.$$

Multiplying by $\frac{100}{54}$, as in the Diagram of Cross-Sections, we have—

☞ $$s = \frac{50c}{54} D. \qquad (19).$$

50. In constructing a diagram to solve this equation, s is laid off on the axis of y, as being the quantity sought; D is laid off on the axis of x, as in the diagrams previously described; and the equation is plotted for successive values of c. The diagram is of so simple a nature that no figure is required.

51. Rule for Use of Diagram.—*Take the vertical line representing the given value of D*—or horizontal dimension—*and follow it up to the inclined line rep-*

resenting the given value of c—or vertical dimension. *Hold the point of intersection with a needle-point, and read off the solidity from the horizontal lines.*

Numerous examples of use are given among the special applications of the diagram below.

It makes no difference, in reality, which dimension is taken as the variable c, and which as the variable D; but it is commonly most convenient to let the vertical lines represent the horizontal dimension, as in the previous diagrams, and the rule is accordingly so given.

52. WHEN THE VALUE OF D IS LESS THAN 8 FEET—which is the smallest shown on the diagram—multiply it by 10, and take one-tenth of the solidity obtained.

WHEN THE VALUE OF D EXCEEDS 88 FEET, OR THAT OF c EXCEEDS 18 FEET, divide that dimension by 2, and take twice the solidity obtained. This is a very simple operation, and the range of the diagram is thus practically extended to sections of 36×176 feet. Greater range is rarely required. With a little practice, double solidities may be read off at once from the horizontal lines; especially if double values be written on them in fine red figures.

Specific Rules for Computation, by the Diagram of Triangular Prisms.

The applications of the Diagram of Triangular Prisms are very numerous. It is sufficient in itself for the computation of all end-area solidities, not only in railroad-work, but also for canals, roads, water-works, etc. Most of these applications are given below, with practical examples of computation.

53. Regular Three-Level Cross-Sections may be computed under the usual rule for numerical computation, as given in paragraph 15. Enter the diagram with the centre-height and sum of distances out; and also with the sum of the side-heights and the half-width of road-bed. The sum of the two quantities obtained is the solidity due to the section.

This rule is especially convenient in computing sections of unusual shape which cannot be taken from the Diagram of Cross-Sections, or for road-beds other than those given among the plates. Sections of 10 or 12 feet centre-height may be taken off at a single inspection, under equation (4), page 15, by first increasing the centre-height by the altitude of the grade-triangle, as given in Table I., page 91, and then deducting the solidity due to it, which is *half* that given in the same table.*

Example.—The section computed on page 23. Follow up the vertical line $D = 32.4$ to the inclined line line $c = 7.4$. Solidity obtained 222. Then follow up line 9.0 to line 9.6 ($5.2 + 4.4$). Solidity obtained 80. Total solidity, $222 + 80 = 302$, as before. Or, enter-

* On account of the fact that the solid to be computed is only 50 feet long instead of 100.

ing with $D = 32.4$; $c = 7.4 + 6.0 = 13.4$, we obtain 402; less 100—the solidity due to the grade-triangle—gives 302 as before.

54. Irregular Three-Level Sections, *or those in which intermediate levels between the centre and side height have been required to give a true expression of the surface,* may be computed by subdividing the section into triangles and trapezoids without plotting, by the method often used in numerical computation, but it is usually deemed best to plot such sections under any circumstances, and the following method is then preferable:

55. Take off the solidity from the Diagram of Cross-Sections, neglecting the intermediates. Then, plot the *surface-lines only* of the cross-section, on ordinary cross-section paper; and draw straight lines from the centre to each side of the section, as shown in Fig. 15,

Fig. 15.

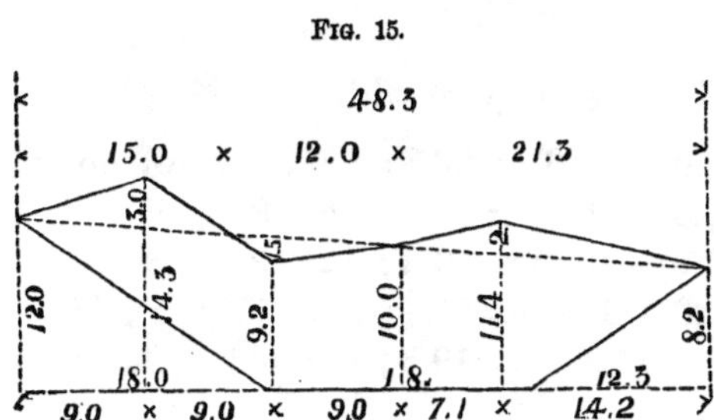

to mark where the surface has been *assumed* to be in taking off the solidity. Compute from the Diagram of Triangular Prisms the areas cut off above and below these lines, and the *algebraic sum* of the quantities obtained gives the correction, plus or minus, as the case may be, to be added to the solidity previously obtained.

EXAMPLE.—The section shown in Fig. 15. First taking off the solidity from the Diagram of Cross-Sections (road-bed 18, 1½ to 1), by entering it with $D=48.3$; $c=10.0$, we obtain 616. This has to be corrected, after plotting the surface-lines, by the solidity due to the three triangles shown in Fig. 15. Entering the Diagram of Triangular Prisms with $D=15.0$, $c=3.0$, we obtain 42; with $D=12.0$, $c=1.5$, we obtain 17; and with $D=21.3$; $c=2.0$, we obtain 40. Then $616+(42-17+40)=681$, which is the solidity due to the section.

The process of computation *without* plotting, by subdividing the section into trapezoids—the triangles outside of the side-slopes being first included and then deducted—is as follows:

Entering the diagram with—

$D=(12.0+14.3)=26.3$;	$c=9.0$;	we obtain	219
$D=(14.3+9.2)=23.5$;	$c=9.0$;	" "	196
$D=(9.2+10.0)=19.2$;	$c=9.0$;	" "	160
$D=(10.0+11.4)=21.4$;	$c=7.1$;	" "	141
$D=(11.4+8.2)=19.6$;	$c=14.2$;	" "	259
			975
Less, $D=18.0$;	$c=12.0$;	$=200$	
$D=12.3$;	$c=8.2$;	$=94$	294
Total solidity, as before,			681

56. In determining the prismoidal correction for such sections, proceed as for regular three-level sections, and neglect the intermediate levels. There is a slight error in so doing, the small solids cut off by the imaginary surface-lines not being computed by the prismoidal formula; but this is commonly in excess, and less than would be involved in reducing to level section; while the latter method *always* tends to a deficiency of solidity.* The correction might be determined by that method, however, as easily, and in the same way, as for regular three-level sections.

57. Five-Level Cross-Sections—i. e., *those in which two additional surface-levels are regularly taken over each road-bed angle*, as shown in Fig. 16—may evidently be divided into three sets of two triangles each, and computed as follows: Enter the Diagram of Triangular Prisms, successively, with the centre-height and width of road-bed, and with the distance out to each slope-stake and corresponding road-bed

* This same tendency to deficiency holds true for *all* transformations of section by an equalizing line, whether it be level or otherwise; and such methods of transformation will commonly give a result more in error than the above simple process, which is probably more *really* correct than if it were mathematically so; since the surfaces of irregular sections are not commonly bounded by straight lines, but by curves to which the straight lines are chords. This question is further considered in Appendix A.

FIG. 16.

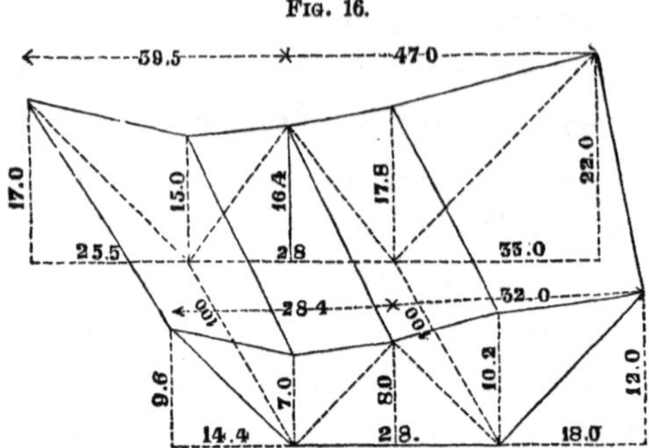

level. The sum of the three quantities obtained is the solidity.*

58. *To determine the prismoidal correction*, a solid of such end-sections may be divided into three prismoids having double triangles for end-sections. The middle one of these, since it has the constant road-bed as a base, is given correctly by end-areas: The error in the other two may be determined by the Diagram of Prismoidal Correction, which (see paragraph 38) applies to all such prismoids, as follows: Subtract the road-bed levels on each side of the centre from each other, and also the distances out to the slope-stakes; and with the

* An *irregular* five-level cross-section—with additional levels taken at a distance $= d$ from those over the road-bed angle—may be taken off without plotting by subtracting $\frac{d}{r}$ from such additional cuttings before computing the section. If the distance d be made equal to the half-width of road-bed, the deductive correction will be the altitude of the grade-prism, as given in Table I., page 91. The process is sufficiently obvious on sketching such a section. A seven-level section, thus computed, requires only five quantities to be taken from the diagram.

differences obtained enter the Diagram of Prismoidal Correction. The sum of the two quantities thus obtained is the total correction. OR, the correction may be approximately determined, at a single operation, by entering the diagram with the differences of *centre* heights and of *total* widths, as if for a three-level section. The error in so doing would commonly be small.

EXAMPLE.—The prismoid shown in Fig. 16. To compute the solidity, entering the Diagram of Triangular Prisms with—

$D = 28.0$, $c = 16.4$,	we obtain	425	
$D = 39.5$, $c = 15.0$,	" "	549	
$D = 47.0$, $c = 17.8$,	" "	775	1,749

For the second section, entering with—

$D = 28.0$, $c = 8.0$,	we obtain	207	
$D = 28.4$, $c = 7.0$,	" "	184	
$D = 32.0$, $c = 10.2$,	" "	302	693
Total end-area solidity,			2,442

Then, to determine the prismoidal correction; entering that diagram with—

$(d - d') = (39.5 - 28.4) = 11.1$;			
$(c - c') = (15.0 - 7.0) = 8.0$,	we obtain	27	
$(d - d') = (47.0 - 32.0) = 15.0$;			
$(c - c') = (17.8 - 10.2) = 7.6$,	" "	35	62
Giving as the *true* solidity,			2,380

Or, approximately, entering that diagram with—

$(D - D') = (86.5 - 60.4) = 26.5,$

$(c - c') = (16.4 - 8.0) = 8.4,$ we obtain 67

Giving, as the net solidity, 2,375

59. Sidings, if made during construction, are usually cross-sectioned by setting independently an additional slope-stake at each original cross-section, making a constant addition, d, Fig. 17, to the original half-width of

Fig. 17.

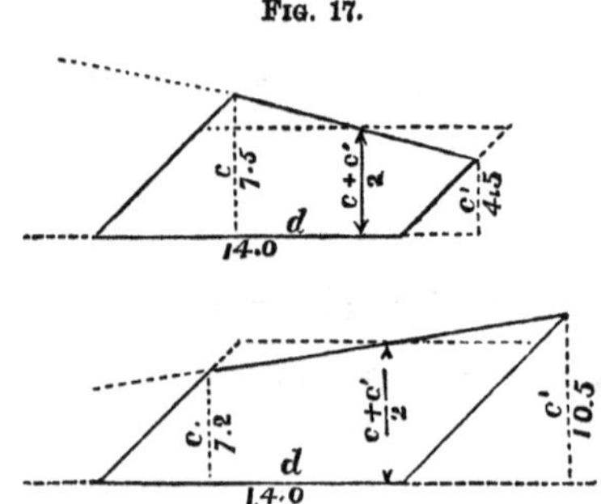

road-bed. The addition thus made to the original section has the form shown in Fig. 17; and, as is evident from the figure, its area $= \frac{c + c'}{2} d$; the section being convertible into a level parallelogram by the dotted equalizing line.

To take off the solidity: Follow up on the Diagram of Triangular Prisms the proper vertical line for d—which always remains the same—till it intersects the inclined line representing the value of $(c + c')$ for each section.

The volume of such prismoids, however great the irregularities of surface, is given correctly by the method of end-areas, as long as the width of road-bed remains the same. When the road-bed also varies, as at the ends of a siding, enter the Diagram of Prismoidal Correction with $(d_1 - d_2)$ and $(c_1 + c_1') - (c_2 + c_2')$; the solid (see paragraph 38) having trapezoidal end-sections.

Example.—The sections shown in Fig. 17. Taking the vertical line $D = 14$, determine its intersection with—

$(c + c') = (7.5 + 4.5) = 12.0$, giving	156
$(c + c') = (7.2 + 10.5) = 17.7$, "	230
Total solidity for a length of 100 feet,	386

60. *Sidings made subsequently to construction* have sections which are only a simpler form of those for borrow-pits, shown in Fig. 19, and can of course be computed in the same way; and without plotting the sections wherever it would be possible in numerical computation. Widening the road-bed for *any* purpose may of course be computed by one of the above methods.

61. Flattening Slopes.—In occasional instances the slope is flattened after cross-sectioning, and it is convenient to compute the added excavation separately from the main body of the cut.

In Fig. 18, let c and $r =$ the original side-height

Fig. 18.

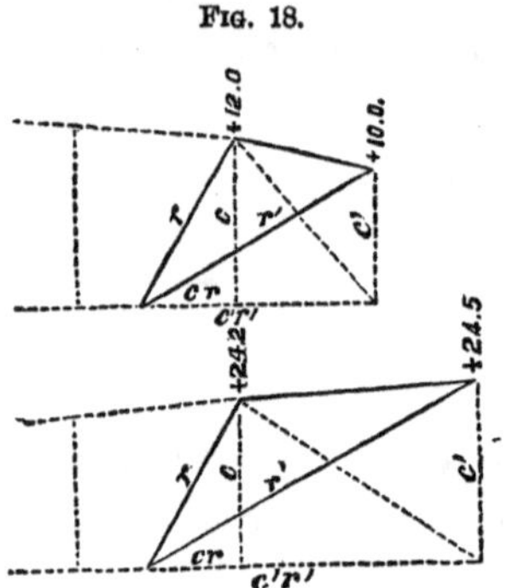

and slope, and c' and $r' =$ the new side-height and slope. Then, as is evident from the figure,

$$A = \frac{cc'r' + c'(c'r' - cr) - c'^2r'}{2},$$

$$= \frac{cc'(r' - r)}{2}.$$

Multiply either the new or old side-height by the difference between the ratios of slope, and consider the product as the variable c; take the other side-height as the variable D, and the solidity due to the section is given by the Diagram of Triangular Prisms.

The prismoidal correction is obtained, when deemed necessary, in the same general manner. Subtract the corresponding side-heights of the end-sections from each other; multiply *one* of these differences by $(r - r')$, and take off the correction from the Diagram of Prismoidal Correction.

EXAMPLE.—The sections shown in Fig. 18; the ratios of slope in the figure being $1\frac{1}{2}$ to 1 and 1 to 1, and the value of $(r - r')$ being consequently $\frac{1}{2}$. Entering the diagram with—

$D = 10.2$; $c = 12.0 \times (r - r') = 6.0$; we have	57
$D = 24.5$; $c = 24.2 \times (r - r') = 12.1$; " "	274
Total end-area solidity,	331

Entering the *Diagram of Prismoidal Correction* with—

$(D - D') = (24.5 - 10.2) = 12.3$;	
$(c - c') = \dfrac{24.2 - 12.0}{2} = 6.1$; we have	72
True solidity for a length of 100 feet,	304

62. BORROW-PITS, which are taken to any regular grade and slope, as for station-grounds, etc., are usually estimated by taking sections similar to Fig. 19, at such

FIG. 19.

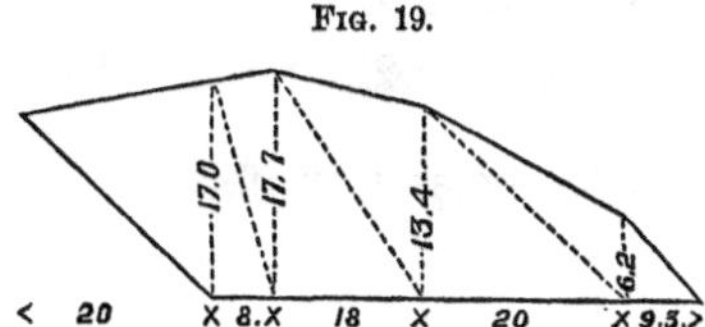

intervals as the ground may require. Such sections, however irregular, can be computed as follows:

4

Divide the section into trapezoids and triangles, by passing vertical lines through each exterior angle, whether at top or bottom. Beginning at one end of the section, consider the length of the first vertical as the variable c, and the horizontal distance between the two verticals on either side of it as the variable D. Take from the Diagram of Triangular Prisms the quantity corresponding to these values of the variables, and perform the same operation for each of the verticals in succession. The *sum* of the quantities thus obtained is the solidity due to the section.

The principle of this process is obvious. It in effect divides the section into triangles, as indicated by the dotted lines in Fig. 19. It is not necessary actually to *draw* the verticals (which are usually some even number of feet apart), nor is it even necessary, in many instances, to plot the section at all.

EXAMPLE.—The section shown in Fig. 19. Entering the diagram with—

$D = 28.0$; $c = 17.0$,	we obtain	441
$D = 26.0$; $c = 17.7$,	" "	427
$D = 38.0$; $c = 13.4$,	" "	471
$D = 29.3$; $c = 6.2$,	" "	168
Total solidity due to the section, for a length of 50 feet,		1,507

63. *The Prismoidal correction for such sections* is difficult to determine exactly, and rarely important;

because, when either the width or average depth of any two sections is identical, or nearly so, end-area solidities are sufficiently correct, however much the other dimensions may vary.

When both the width and average depth vary importantly, the following gives a close approximation:

Determine the difference between the width and between the average depth of the two sections; take off the correction from the Diagram of Prismoidal Correction; *multiply it by* 2; and subtract it from the end-area solidity: UNLESS the deepest cut is at the narrowest end of the solid; in which case *add* it to the end-area solidity.

64. IRREGULAR AND EXTENSIVE BORROW-PITS are usually laid off in rectangles, sometimes of not more than ten feet square. The volume of each rectangular prism will be given by entering the diagram with *the sum of the four edges and the area of the surface divided by* 100.*

EXAMPLE.—A rectangle 25 feet square, and the four corner cuttings, + 6.8 + 8.6 + 10.8 + 9.6, = 35.8.

* The volume of *triangular* prisms is given on the Diagram of Prismoidal Correction, by entering with the sum of the *three* edges and the area divided by 100; after doubling the vertical scale, as suggested in paragraph 74. Or, the volume may be taken off from the Diagram of Triangular Prisms, by entering it with the sum of the three edges and the area divided by 100, and increasing the quantity obtained *by one-third.*

Entering the diagram with—

$$D = 35.8\,;\ c = \frac{625}{100} = 6.25,\ \text{we obtain } 208.$$

This method has no special advantage over numerical computation for a continuous series of such solids, but there are nearly always more irregular solids adjacent to and intermingled with them. For this reason, and to enable the subsequent levels to be taken wherever the new surface may require, such notes are often plotted *as sections*, and may then be computed as already described in paragraph 62. Fig. 20 shows such

Fig. 20.

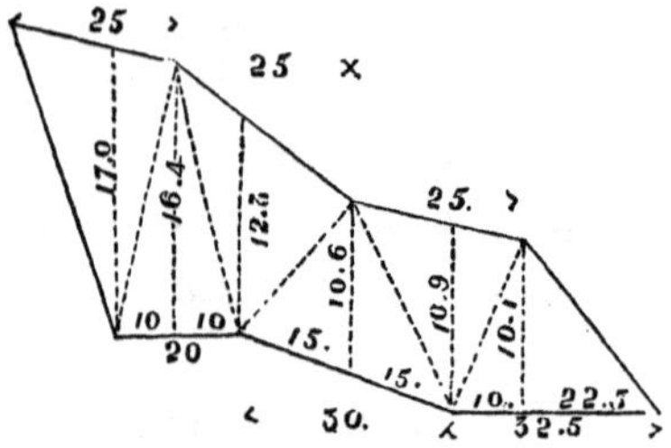

a section. The solidity due to it, as taken off from the diagram, was 1,834 cubic yards.

65. Ditches may be computed as follows: Take the sum of the top and bottom widths as the variable D, and the depth as the variable c, and enter the diagram.

Example.—The sections shown in Fig. 21.

Entering the diagram with—

$D=15.3$; $c=4.6$, we obtain	65
$D=24.2$; $c=9.2$, " "	210
Total solidity for a length of 100 feet,	275

FIG. 21.

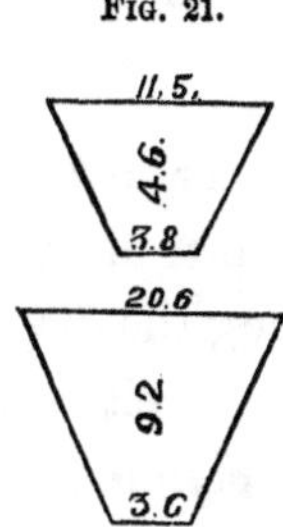

The error in end-area solidities in ditches is generally too small for consideration. The Diagram of Prismoidal Correction, however, applies to solids of trapezoidal end-sections (see paragraph 38), by entering it with the difference between the centre-heights and between the sum of top and bottom widths. For the solid in Fig. 21, the correction thus obtained is 14 cubic yards.

66. CROSSINGS—or an excavation to enable a highway to cross a railway at grade—are, in general, simply two irregular wedges—frequently oblique—placed back to back, with a common base. Enter the diagram with the sectional dimensions of the base or mid-section (viz., the depth and the sum of the top and bottom widths), and multiply the quantity obtained by the extreme length—at right angles to the mid-section—of the solid, regarded as a fractional station.

This rule is of course not strictly correct, but such solids are commonly small and irregular. The Diagram of Prismoidal Correction furnishes an exact process, by paragraph 38.

67. FOUNDATION-PITS may always be regarded as made up of one or more vertical prisms. The volume of each is given by entering the diagram with the sum of the four vertical edges and either one of the horizontal dimensions, and multiplying the quantity obtained by the other, regarded as a fractional station.* The sides are not always vertical, but their top and bottom dimensions are commonly averaged. The exact point at which to average is at .4 of the depth, or $\frac{d}{\sqrt{3}}$ from the bottom.

EXAMPLE.—A culvert-pit, 25 feet long on the upper side, 38 feet long on the lower side, and 15 feet wide. Cuttings at the upper end, 2.0 and 3.2, at centre, 4.6 and 3.8, at lower end, 2.8 and 2.2.

Entering the diagram with—

D=25.0; c=(2.0+3.2+4.6+3.8)=13.6, we obtain 315

D=38.0; c=(2.8+2.2+4.6+3.8)=13.4, " " 472

787

Multiplied by .15 = 78.7 + 39.3 = 118 cubic yards.

* OR, by paragraph 65, by entering with the sum of the four edges and *the area divided by* 100.

We have thus seen that every form of irregular earthwork may be computed from the diagram, and the reader will find it an advantage to do so invariably. On some solids the saving of labor may seem to be small, and if it be attempted to take irregular dimensions directly from field-books or plotted cross-sections, the process is too slow to gain the full advantage of the diagrams; but if forms be made up at once for the entire estimate, as shown in paragraph 118—which most engineers deem necessary under any circumstances—the actual computation is reduced to simple addition and a few trifling multiplications. To prepare the tabular forms necessarily requires some judgment and care, but the number of quantities to be read off is then of very small importance, and the *main* advantage of the process is the greater certainty of correctness, which experience will soon demonstrate.

Art. IV.—*Computation by Henck's Formula.*

68. The method of computation for thorough-cut sections given in Henck's "Field-book for Railroad Engineers," is stated, on page 100, to be based on the assumption that "the surface on each side of the centre-line may be considered to be divided into two triangular planes by a diagonal passing from one of the centre-heights to one of the side-heights. A ridge or depression will, in general, determine which diagonal ought to be taken as the dividing line, and this diagonal must be noted in the field. If the surface happen to be a plane, or nearly so, the diagonal may be taken in either direction."

On any given solid, the solidity given by the prismoidal formula—under the assumption as to surface stated in paragraphs 33 and 34—is precisely a mean between those obtained under Henck's formula, by first running each diagonal in either direction, and then reversing them. This is easily proved by comparing the formulæ; * but a demonstration is omitted, as the tabular example given toward the end of paragraph 117, shows both the nature and amount of the difference between them. It will be seen that on ordinary solids such differences are small; and, inasmuch as a natural surface which is not a plane is never exactly divisible into two triangular planes, it may be questioned if either method is more accurate than the other. The office computations, however, are about equally simple, and in the field-work there is only the slight additional labor of noting the direction of the diagonal.

Diagram for Computation by Henck's Formula.

69. The formula of solidity by this method is obtained by subdividing the solid into pyramids; and, as

* As has been done by Prof. Gillespie. See "Roads and Railroads," revised edition, p. 378.

given on page 103 of Henck's "Field-book," is as follows; the demonstration being omitted, as the volume is accessible to all:—

$$S = \Big[(d_1 + d'_1)\,c_1 + (d_2 + d'_2)\,c_2 + DC + D'C'$$
$$+ \frac{w}{2}(h_1 + h_2 + H + h_1' + h_2' + H')\Big]\frac{l}{6},$$

the nomenclature being the same as in Fig. 4, page 14, except that the *capital* letters in this case designate dimensions touched by diagonals.

To reduce the number of variables, we know that in all cases $h = \dfrac{d - \frac{w}{2}}{r}$; and, substituting this value, we have—

$$S = \Big[(d_1 + d_1')\,c_1 + (d_2 + d_2')\,c_2 + DC + D'C'$$
$$+ \frac{w}{2r}(d_1 + d_1' + d_2 + d_2' + D + D') - \frac{6w^2}{4r}\Big]\frac{l}{6}. \quad (20).$$

This equation can now be subdivided into the four following equations of symmetrical form:

$$\left.\begin{aligned}
S = S_1 &= \left\{\left(c_1 + \frac{w}{2r}\right)(d_1 + d_1') - \frac{3w^2}{8r}\right\}\frac{l}{6}\\
+ S_1' &= \left\{\left(c_2 + \frac{w}{2r}\right)(d_2 + d_2') - \frac{3w^2}{8r}\right\}\frac{l}{6}\\
+ S_2 &= \left\{\left(C + \frac{w}{2r}\right)\quad D \quad - \frac{3w^2}{8r}\right\}\frac{l}{6}\\
+ S_3 &= \left\{\left(C' + \frac{w}{2r}\right)\quad D' \quad - \frac{3w^2}{8r}\right\}\frac{l}{6}
\end{aligned}\right\} (21).$$

Substituting 100 for l, and dividing the equation by 27, we have—

$$\left.\begin{aligned} S_1 &= \frac{100}{162}\left(c_1 + \frac{w}{2r}\right)(d_1 + d_1') - \frac{25w^2}{108r} \\ S_1' &= \frac{100}{162}\left(c_2 + \frac{w}{2r}\right)(d_2 + d_2') - \frac{25w^2}{108r} \end{aligned}\right\} \quad (22).$$

☞

$$\left.\begin{aligned} S_2 &= \frac{100}{162}\left(C + \frac{w}{2r}\right) D - \frac{25w^2}{108r} \\ S_3 &= \frac{100}{162}\left(C' + \frac{w}{2r}\right) D' - \frac{25w^2}{108r} \end{aligned}\right\} \quad (23).$$

$$S = S_1 + S_1' + S_2 + S_3. \qquad (24).$$

The constants in the four equations (22) and (23) are all identical, and hence they may all be solved on a single diagram.

70. The general method of construction and the arrangement of the variables for this diagram is the same as shown in Fig. 5, for the Diagram of Cross-Sections. The horizontal dimension, $(d + d')$, equations (22), or D, equations (23), is taken as the variable x; the solidity, S_1, etc., as the variable y, and the centre-height, c, is treated as a constant, and hence values of it are represented by inclined lines. The appearance of the diagram (Plate X.) is somewhat peculiar, owing to the fact that quantities are to be taken off, not only with the centre-height and total width, from slope-stake to slope-stake, of each section, but also with the centre-height and the distance out from the *centre-line* to a slope-stake at the other extremity of a diagonal. Hence a much greater range is required in the values of D given for each centre-height, and it will be observed that 30 feet centre-heights are shown in each of the two lower subdivisions, and that the diagram begins at $D = 9$ feet, instead of 18 feet, as in the Diagram of Cross-Sections. In the two upper subdivisions, however, the values of D become too great to ever occur in equation (23), as the distances out to a single slope-stake, and hence only a limited range along the Curve of Level Section is given, to enable the quantities S_1 in equations (22) to be taken off.

The Curve of Level Section is of no special advantage, except to serve as a check against gross errors in the notes, as explained in paragraph 25. It enables any section to be reduced to an equivalent level centre-height, however, by the method of paragraph 24.

71. RULE FOR USE OF DIAGRAM.—*Enter the diagram with the centre-height and the horizontal distance from slope-stake to slope-stake, and take off the quantity S_1 for each section. Also, enter with the dimensions touched by diagonals, and take off the quantities S_2 and S_3 for each solid.*

TO TAKE OFF QUANTITIES: *Follow up the vertical line representing the given horizontal dimension to the inclined line representing the given centre-height. Hold the point of intersection, and read off the solidity from the horizontal lines.*

By "the dimensions touched by diagonals" is meant the centre-height at one extremity of a diagonal, and the distance out to the slope-stake at the other extremity. The side-heights do not enter into the calculation, being merely functions of the distances out.

Example.—The same as given on page 103 of "Henck's Fieldbook." Entering the diagram (Plate X.) with—

c=13.6 ; $(d+d')$=45.0, the quantity taken off is 494 (S_1)
c= 8.0 ; $(d+d')$=42.0, " " " " " 313 (S_1')
C=13.6; D =27.0, " " " " " 277 (S_2)
C= 8.0; D =21.0, " " " " " 131 (S_3)

Total, 1,215

Or, reversing the direction of the right-hand diagonal, we obtain the first two quantities, and the last one, as before, equal to 938
and entering the diagram with
C=8.0; D=24.0, the quantity taken off is 157 (S_2)

Total, 1,095 cu. yds.

As given in the volume, the first quantity is 32,820 cubic feet, equal to 1,215.56 cubic yards; and the second (given on page 110) is 29,600 cubic feet, equal to 1,096.3 cubic yards. The difference is the error of observation.

By referring to the lower left-hand corner of the diagram, it will be seen that, in rare instances, the equations S_2 and S_3, equations (23), may have a *minus* value, when the value of D is very small. It occurs so seldom, however, as only to require mention. It is due to the equal distribution of the quantity $\frac{6w^2}{4r}$, equation (20), (which is equal to the volume of the grade-prism), among the four succeeding equations (21), in order that they may be symmetrical, and hence solvable on the same diagram.

A tabular example of computation by this method is given in Chapter IV., paragraph 117.

72. Sections of over 30 feet centre-height may be taken off as follows: Subtract from the centre-height the altitude of the grade-prism, 6.0 for Plate X. Then, divide both the centre-height and the horizontal dimension by two, and take off a quantity from the diagram. Multiply it by four, and add to the product the quantity 150, equal to three times the value of the final term $\frac{25w^2}{108r}$, equations (22) and (23), or three-quarters of the volume of the grade-prism for a length of 100 feet.

This rule applies alike to the quantities S_1, S_2, and S_3.

The *rationale* of this process is similar to that given in paragraph 22 for the Diagram of Cross-Sections. The centre-height, as obtained above, is equal to half the centre-height of the section if it included the grade-triangle, *less the altitude* of that triangle. Hence, four times the quantity taken off, plus four times that due to the area of the grade-triangle—or the value of the final term $\frac{25w^2}{108r}$, equations (22) and (23)—will give the quantity required if the grade-triangle were *included* in the section; and four times the quantity taken off, plus *three* times the value of the final term, gives the quantity required.

73. VERY IRREGULAR SECTIONS OF FROM 25 TO 30 FEET CENTRE-HEIGHTS may sometimes exceed the limits of the diagram, and may be taken off as follows: Before entering the diagram—

Subtract from the centre-height,	And add to the quantity taken off,
1.6 (exactly, 1.62)	D
or	
3.2 (" 3.24)	$2D$
or	
8.1	$5D$

The *rationale* of this process is sufficiently obvious from paragraph 23, giving a similar one for the Diagram of Cross-Sections.

This process may also be used to take off sections from 30 to 35 feet centre-heights, instead of that given in the preceding paragraph. It is slightly more accurate, since the error of observation is not multiplied by four. Also, the method in the preceding paragraph is of course applicable in all cases where a section cannot be directly taken off. Sections with centre-heights of less than 25 feet, however, can always be taken off directly, however irregular.

74. SIDE-HILL EARTHWORK may be computed by the method of "Henck's Fieldbook" from the Diagram of Prismoidal Correction, by assigning values to the horizontal lines equal to double those shown, thus making the scale identical with the diagrams for thorough-cut sections. The formula of solidity is given on page 107 of that volume, and need not be repeated.

Enter the diagram with the side-height and the width of excavation at the road-bed, for each section; and with the dimensions touched by the diagonal, for each solid. The sum of the three quantities obtained gives the volume of the solid.

The "dimensions touched by diagonals" are the side-height at one end of it and the width of excavation at the road-bed at the other end.

EXAMPLE.—The following cut taken from page 107.

NOTES FROM VOLUME.			SOLIDITIES TAKEN OFF IN CUBIC YARDS.		
Station.	Road-bed.	Side-height.	Quantities for each Section.	Quantities for Diagonals.	Volume of each Solid.
0	2.0	1.0	1		
1	8.0	6.0	30	5	36
2	10.0	7.0	43	34	107
3	13.0	7.0	56	43	142
4	9.0	4.0	22	32	110

Total, 152 − (1 + 22) + 152 + 114 = 395 c.y.

As given in the volume, 10,700 cu. ft. = 396.3 c.y.

75. Irregular Earthwork may also be computed by the method of "Henck's Fieldbook" from the Diagram of Prismoidal Correction, after doubling the scale as above. The formula of solidity (p. 109) is based upon a subdivision of the solid into vertical triangular prisms —the solid bounded by the plane of the road-bed extended, the plane of the side-slopes, and a vertical plane passing through the two slope-stakes on each side, being first included and then deducted. The horizontal length of each of these prisms is equal to the length of the solid, the width is a given horizontal dimension, and the altitude at each vertical edge is a recorded cutting.

Enter the diagram, for each prism into which the solid is subdivided, with the given horizontal dimension and the sum of the three given cuttings. For the deductive solid: Enter the diagram with *either one* of the side-heights and the sum of both; and with the *other* side-height as both horizontal and vertical dimension. Add the two quantities obtained, multiply their sum by the ratio of slope, and subtract the product from the sum of the quantities obtained above.

Example.—The solid given on page 109 of "Henck's Fieldbook." Entering the diagram with—

$D = 17.0$; $c = 4.0$,	the quantity obtained is	42
$D = 18.0$; $c = 9.0$,	" " " "	100
$D = 23.0$; $c = 5.0$,	" " " "	71
$D = 24.0$; $c = 12.0$,	" " " "	178
$D = 21.0$; $c = 9.0$,	" " " "	117
Total,		508

LESS: $D = (6 + 8)$; $c = 6$, giving	52	
$D = 8$; $c = 8$, "	39	
Total deductions,	$91 \times \frac{3}{2}$	$= 136.5$
Net total,		371.5

Multiplied by .50, for a length of 50 feet, gives 185.75
As given in the volume, 5,000 cu. ft., equal to 185.19

WHEN A NUMBER OF IRREGULAR SOLIDS SUCCEED EACH OTHER, whether of equal length or not, the following mechanical process, equivalent to the above, may be used:

For each section, enter the diagram with each cutting in succession, and the horizontal distance between the cuttings on either side (for the side-heights there is, of course, a cutting only on *one* side). From the sum of the quantities obtained, deduct that given by entering the diagram with the sum of the side-heights and the sum multiplied by r.

Also, *for each solid*, enter the diagram with the cutting at the apex of each triangular surface-plane and with the length of the base of the triangle in the other end-section. From the sum of the quantities obtained, deduct that given by entering the diagram with the sum of the side-heights of one end-section, and the sum multiplied by r, of the other.

Then the total volume of each solid, for a length of 100 feet, is given by adding together the quantities for the end-sections and for the solid.

When one end-section is three-level, as must frequently be the case, the quantity for each section obtained as above is equal to the quantity S_1, equation (22), $+ 16.7$, the difference between $\frac{1}{4}$ of the solidity of the grade-prism (50 yards), which is *actually* deducted in equation (21), and $\frac{1}{3}$ of that solidity (66.7 yards), which strict theory requires to be deducted. Also, when Plate X. is in use, by the rule of paragraph 109, for another road-bed than 18 feet, the quantity S_1 as taken off must be corrected by *two-thirds* only of the correction in Table II. (plus 16.7 as before) to have it equal to the quantity obtained as above: For example, if the road-bed be 20 feet, the correction in Table II. is $- 70.4$; $\times \frac{2}{3} = - 46.9$; $+ 16.7 = - 30.2$.

Art. V.—*Computing Excavation on Curves.*

76. Although the error due to curvature is not commonly of very great moment, it will sometimes reach large dimensions, for the amount of which the following gives some basis of comparison. With a 50 feet centre-height and 15° surface-slope on a 10° curve it amounts to 4.65 per cent., or 965 cubic yards; with a centre-height of 25 feet, and the same surface-slope, it amounts to 2.6 per cent., or 164 cubic yards. With a centre-height of 10 feet it reduces to 22½ cubic yards. The *percentage* of error is directly as the degree of curvature, and as the centre-height measured from the intersection of the side-slopes prolonged. The *amount* of error is as the degree of the curve and square of the centre-height, and, approximately, as the square of the surface-slope. It is not diminished by increasing the frequency of sections, as is quite frequently assumed, and indeed is stated by a recent writer.* This becomes sufficiently evident without demonstration by considering the case of any one given solid. Plainly, the error, *due to curvature*, is only *distributed* by interposing an additional section, since the angle between the end-planes of the original solid remains the same.

Diagram of Correction on Curves.

77. As given in Henck's "Fieldbook for Railroad Engineers," page 112, the correction for curvature due to any section, in cubic feet, for a full station, is as follows:

$$C = \left[\frac{c}{2}(d - d') + \frac{w}{4}(h - h')\right] \frac{100\,(d + d')}{3R};$$

R being equal to the radius of curvature, and the signification of the letters otherwise the same as in Fig. 4.

Substituting for h its value, $\dfrac{d - \frac{w}{2}}{r}$, we have—

$$C = \left(\frac{c}{2} + \frac{w}{4r}\right)(d - d')\frac{100\,(d + d')}{3R}.$$

* "Easy Rules for the Measurement of Earthworks," by Ellwood Morris, C. E., p. 81

Substituting, also, l for 100, to make the formula of general application, we have—

$$C = l\left(\frac{c}{2} + \frac{w}{4r}\right)(d + d')\frac{d - d'}{3R}. \qquad (26).$$

From equation (4), we have as the solidity of a *right-lined prism* of any section—

$$S = l\left(\frac{c}{2} + \frac{w}{4r}\right)(d + d') - \frac{lw^2}{4r}. \qquad (27).$$

Then, if the same solid be considered as having a *curved* centre-line, we have for its solidity, by combining equations (26) and (27)—

$$S \pm C = l\left(\frac{c}{2} + \frac{w}{4r}\right)(d + d')\left(1 \pm \frac{d - d'}{3R}\right) - \frac{lw^2}{4r} \quad (28),$$

the character of the sign, plus or minus, depending on which side of the curve lies the greatest value of d.

78. The only difference between equations (27) and (28) is that in the latter the value of $(d + d')$, or D, is to be multiplied by a certain coefficient. Manifestly, then, the correction may be applied to the value of D for each section before computing areas or solidities, rendering unnecessary any consideration of the length or volume of the solid. Letting $C_D =$ the amount of change to be made in each value of $(d + d')$, or D, and substituting 573, the radius of a 10° curve, for R, we have—

☞ $$C_D = \frac{d - d'}{1,719} D.^* \qquad (29).$$

79. For *Side-Hill* sections we might obtain in a similar way, as close approximations; for the correction for the distance D_1, Fig. 10—

$$C_D = \frac{D_1 - x}{1,719} D_1, \qquad (30);$$

and similarly for D_2—

$$C_D = \frac{D_2 + x}{1,719} D_2, \qquad (31);$$

both of which may plainly be taken off from the Diagram of Correction on Curves, when required, but the correction on such sections is commonly very small. Equation (29) is equally applicable to computation by Henck's formula, except that d and d' successively must be taken as the variable D.

* For *numerical* computation, equation (29) would take the still simpler form for a 1° curve—

$$C_D = \frac{7}{12} \frac{d^2 - d'^2}{10,000}.$$

With a table of squares this requires hardly any computation. Thus, take the following section, on a 6° curve—

$$\frac{99.0}{+\ 60.0} \qquad +\ 40.0 \qquad \frac{48.0}{+\ 26.0}$$

$99^2 - 48^2 = 7,497$; $\div\ 10,000 = .75 \times \frac{7}{2} = 2.6$, the required correction. Then $147.0 - 2.6 = 144.4$, the corrected width on top, which rectifies the solid. Area of section by equation (4), from corrected notes, 3,267; from original notes, 3,327; error, 60; nearly 2 per cent.

80. In construction, D is taken as the variable x, and C_D as the variable y, following the general arrangement of all the previous diagrams, and the equation is plotted for successive values of $(d - d')$.

81. The formula of correction may easily be modified to give the correction in cubic yards for a full station, but the form given would seem to be most convenient for general use. If desired, however, the correction in cubic yards may be taken off from the same diagram.

Referring to equation (28), it is evident that the coefficient $\left(1 \pm \frac{d - d'}{3R}\right)$ may be regarded either as multiplying the adjacent factor $(d + d')$ or the entire term, which is equal to the volume of the solid *plus* that of the grade-prism, $\frac{lw^2}{4r}$. If the latter be neglected as a comparatively small quantity,* we have from equation (26), letting $S =$ the solidity of the solid, as previously computed, and substituting 573, the radius of a 10° curve for R—

☞ $$C = \frac{d - d'}{1{,}719} S. \qquad (32).$$

This equation has the same constant as equation (29); consequently the same diagram applies, but as the values of the variables are commonly much greater in equation (32) than in equation (29), it is necessary to conceive both the vertical and horizontal scale to be

* Or its volume as given in Table I. may be added to S.

multiplied by 100. The horizontal lines marked 1.0, 2.0, etc., are then equal to 100 and 200 cubic yards, and the vertical lines marked 10, 20, etc., are then equal to 1,000 and 2,000 cubic yards.

82. Rule for Use of Diagram: To rectify the sectional dimensions of the solid. *Take the vertical line representing the* sum *of the distances out,* $(d+d')$ *or* D, *for the given section, and follow it up to the inclined line representing the* difference *between the distances out, or* $(d-d')$. *The horizontal line passing through the point of intersection gives the correction required for a* 10° *curve.*

For other degrees of curvature, multiply by the proper decimal of 10°, and correct the recorded value of D before taking off solidities from the Diagram of Cross-Sections.

The correction is to be *added*, when the highest ground is on the convex side of the curve; and *subtracted*, when the highest ground is on the concave side. At a tangent point, take half the correction.

In determining the prismoidal correction, use the corrected width on top. The solidity then obtained is absolutely correct, while the original formula involves the same percentage of error as the end-area solidity.

Example.—The prismoid shown in Fig. 8, supposed to be on a 10° curve. Entering the diagram of correction with—

$D=55.5$; $(d-d')=11.1$; we obtain .4
$D=45.0$; $(d-d')=\ \ 8.4$; " " .2.

Then $55.5 - .4 = 55.1$, and $45.0 - 0.2 = 44.8$; which are the corrected values of D with which to compute the section. Entering, then, the Diagram of Cross-Sections with—

$D = 55.1$; $c = 9.8$; we should obtain	706
$D = 44.8$; $c = 8.4$; " " "	497
Solidity for the curved prismoid,	1,203

As previously computed for a right-lined solid, the solidity was 1212. The difference is of course small for such small and regular sections. The highest ground was here assumed to be on the concave side of the curve. Had it been assumed on the other side, the solidity obtained would have been 1221.

83. To USE THE DIAGRAM TO TAKE OFF CORRECTIONS IN CUBIC YARDS, *take the vertical line representing the given solidity, follow it up to the inclined line representing the given value of* $(d - d')$, *and read off the correction; the values of both the vertical and horizontal lines which are marked on the plate being supposed to be multiplied by* 100.

Multiply the corrections obtained by the degree of the curve, regarded as the decimal of 10°.

The corrections may be thus determined, either for each diagram quantity before adding them together, or for the solidity as finally computed; but in the latter case the average of the values of $(d - d')$ for the two end-sections must be taken.

Example.—The prismoid shown in Fig. 8, as before. To correct each diagram quantity separately—entering the diagram with—

$S = 712$; $(d - d') = 11.1$;	we obtain	6	cu. yd.
$S = 500$; $(d - d') = 8.4$;	"	3	"
Total correction,		9	"

Or—to determine the correction for the entire solid at once—entering the diagram with—

$S = 1{,}212$; $(d - d') = \dfrac{11.1 + 8.4}{2} = 9.75$; we obtain 9 cubic yards.

84. In Sidings, Borrow-Pits, and Double-Track Excavation, the *percentage* of error due to curvature is considerably increased, owing to the fact that the whole excavation lies on one side of the centre. A formula of correction for it is easily obtained, but it was not deemed of sufficient importance to include a diagram among the plates. The error in such work is nearly as the extreme distance out from the centre, and when that distance is 100 feet it amounts to nearly one per cent. per degree of curvature.

85. Fig. 21 shows the general form of sections of this class. On curved work the prism due to that section, as computed from the Diagram of Trianglar Prisms, may be regarded as a *solid of revolution*, the solidity of which is equal to the area of the section multiplied by the distance passed over by its centre of gravity. The error is caused by the fact that this centre of gravity lies at a distance, x, from the centre-line on which the

Fig. 22.

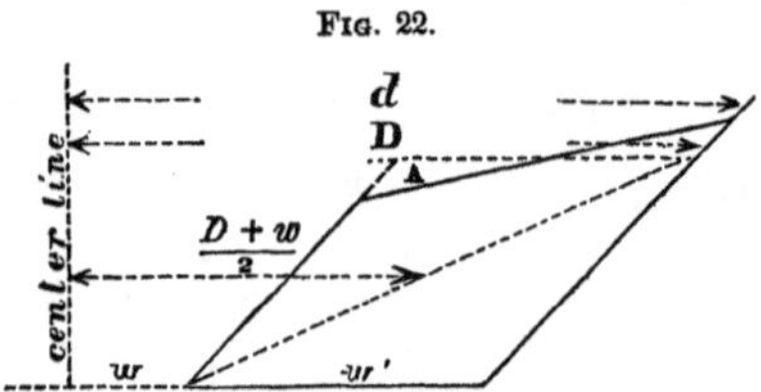

length of the solid was measured, causing a variation in the length of its path from the recorded length. If the radius be R, $\frac{x}{R}$ will be the coefficient of this error, and, if multiplied by the solidity, will give the correction.

To determine x, transform the section into a rhomboid, by the dotted equalizing line shown in the figure. By so doing the small triangle A has its centre of gravity moved a distance of $\frac{2}{3}w'$, but the error is of little moment, and tends to diminish the correction.

The centre of gravity is now the middle point of either diagonal, and, without special demonstration, we have from Fig. 21, as its horizontal distance, x, from the centre-line—

$$x = \frac{D + w}{2}.$$

Substituting this value of x, and 573 for R, and multiplying by the solidity, s, as taken off from the Diagram of Triangular Prisms, we have—

☞ $$C = \frac{D + w}{1,146} s. \qquad (33).$$

86. A little diagram, similar to the one given, is easily constructed to solve this formula, but corrections under it may also be taken off as follows from the diagram given on Plate IX., if the scale be supposed to be multiplied by 100, as suggested in paragraph 81:

Comparing equations (32) and (33), it will be observed that the former is equal to $\frac{2}{3}$ of the latter, for any given values of the variables. Consequently, to take off corrections under equation (33), enter the diagram with the given values of D and s, and multiply the correction obtained by once *and a half* the degree of the curve—regarded as the decimal of 10° as before. Thus, on a 4° curve, multiply by six-tenths.

87. The distance D in this equation is not identical with d, the distance out to the slope-stake, but when the surface-slope is not over 12° or 14° it may be considered so without appreciable error, by dropping the final decimals. For steeper slopes, a very close approximation is given by correcting d by half the difference between the two side-heights. D has also to be increased, in every case, by the constant quantity w (which has sometimes a minus value). The corrections, after being multiplied by the degree of the curve as above, are added to, or subtracted from, the solidity for each section taken off from the Diagram of Triangular Prisms.

This formula may be applied with sufficient accuracy to wider excavations, with irregular surfaces, as in grading for station-grounds, etc., when so cross-sectioned as to require it.

Art. VI.—*Computation of Preliminary Estimates.*

88. Preliminary Estimates may be classified, according to the accuracy attempted, as being made—

First. From centre-heights only.

Second. From centre-heights and a single surface-slope, assumed to be uniform.

Third. From centre-heights and two or more surface-slopes for each section.

The surface-slopes will be assumed to be given in degrees. If given by the rise or fall in some even distance, as ten feet, they are readily turned into degrees by a table of natural tangents.*

Computations by any one or all of these methods may be made, on a single plate, for all road-beds having the same side-slope. Before taking up that diagram, however, we will consider the Diagram of Slope-Stakes, as its equation will be subsequently required.

* When surface-slopes are taken with rods, they may be read off in degrees by laying off natural tangents on the vertical rod instead of feet and tenths.

Diagram of Slope-Stakes.

89. In any section with two surface-slopes given and lettered as in Fig. 23, let S and R = the angle of surface and of the side-slopes with the horizon, and r = the ratio of slope = cot R. Then, when the surface slopes *away from* the road-bed, we have—

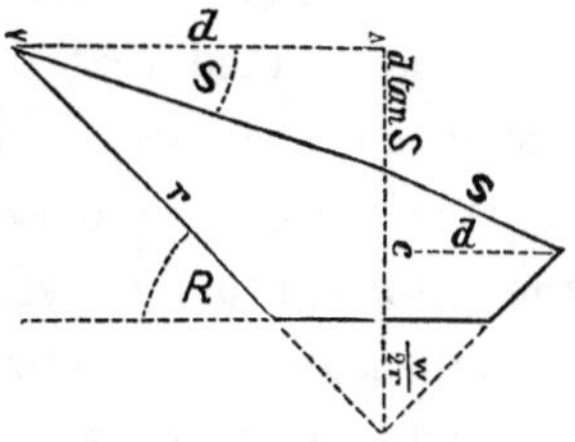

FIG. 23.

$$c+\frac{w}{2r}+d\tan S=d\tan R,\ \text{and}$$

$$d=\frac{c+\frac{w}{2r}}{\tan R-\tan S}. \qquad (34).$$

When the surface slopes *toward* the road-bed, we have—

$$c+\frac{w}{2r}-d\tan S=d\tan R,\ \text{and}$$

$$d=\frac{c+\frac{w}{2r}}{\tan R+\tan S}. \qquad (35).$$

90. From these equations the Diagram of Slope-Stakes on Plate IX. is constructed, the surface-slope S

being taken as the constant and c as the variable y. It affords, in connection with the Diagram of Cross-Sections, a very *accurate* method of making preliminary estimates, though it is a less concise and convenient one than that given hereafter. The diagram is also designed, however, for setting slope-stakes for construction, in work where it is deemed that the field-notes can be taken with sufficient exactness. Much labor is saved over the ordinary process of cross-sectioning by the level, though the latter is usually deemed most satisfactory. The single diagram given is readily applied to all road-beds with 1½ to 1 side-slopes, by the method of paragraph 110, Chapter III.

91. Rule for Use of Diagram.—*Follow up the vertical line representing the given centre-height, to the inclined line representing the given surface-slope. The distance out to where the slope-stake would be is then read off from the horizontal lines.*

If the point of intersection falls below the horizontal line $d = \frac{w}{2}$, *it indicates that the section is side-hill.*

Example.—24 feet centre-height; surface-slopes, *left*, + 15°, *right*, − 12°.* Entering the diagram with—

* The surface-slopes are marked *plus* and *minus*, according as the surface is inclined *away from* or *toward the road-bed*, and not according to whether the ground slopes up-hill or down-hill. In excavation the two are coincident, but in embankment an up-hill surface-slope, being toward the road-bed, must be regarded as minus.

$$c = 24\,;\ d = \begin{cases} +15^\circ;\ \text{we have } 62.8 \\ -12^\circ;\ \text{“} \quad \text{“} \quad 28.4 \end{cases} \text{total } 91.2.$$

DISTANCES OUT FOR SIDE-HILL SECTIONS *are given in the same manner, by that part of the diagram to the left of the vertical axis, the diagram being entered with a minus centre-height.* For an explanation, see paragraph 131 and Fig. 24.

FIG. 24.

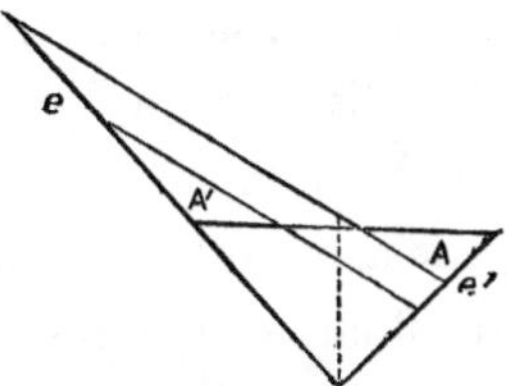

EXAMPLE.—Centre-height, – 1.0, i. e. (for a *fill* road-bed), *cut* 1.0, surface-slope $+15^\circ$; then $d = 7.5$.

Diagram for Sections of Double Slope.

92. To determine at a single operation the solidity due to the half-section on either side of the centre, Fig. 23; if we let $c + \frac{w}{2r} = C =$ the centre-height measured from the intersection of the side-slopes prolonged, we have—

$$A = \frac{Cd}{2} - \frac{w^2}{8r}.$$

But by equations (34) and (35) $d = \frac{C}{(\tan R \pm \tan S)}$.

Substituting this value, and multiplying by $\frac{100}{27}$, to give the solidity of the half-section for a full station, we have—

$$☞ S = \begin{cases} S_1 = \frac{100}{54} \frac{C^2}{(\tan R - \tan S)} - \frac{25w^2}{54r}, & (36). \\ S_2 = \frac{100}{54} \frac{C^2}{(\tan R + \tan S)} - \frac{25w^2}{54r}. & (37). \end{cases}$$

These equations are of the second degree, and their loci are parabolas. The Diagram for Sections of Double Slopes is constructed from them, the surface-slope S being taken as the constant and C as the variable y. The value of w in the final term is taken as zero, in construction, thus including the grade-prism in the section.

93. Rule for Use of Diagram.—Increase all the centre-heights of the sections to be computed by the constant $\frac{w}{2r}$, as given in Table I., page 91. Then, *follow up the vertical line representing the centre-height, to the inclined line representing the surface-slope to the right and to the left successively, and read off from the horizontal lines the solidity due to each half-section;* which may be recorded either separately or in sum total.

The solidity of the grade-prism has then to be subtracted, which may be done for each solid separately or for the whole length of the cut. A table for this purpose is given on page 91.

EXAMPLE.—24.4 feet centre-height, road-bed 12 feet, surface-slopes $+15°$, $-12°$. $24.4 + 4.$ (as given in Table I.) $= 28.4$. Entering the diagram with $c = 28.4$, $S = +15°$ and $-12°$, we have $3,750 + 1,700 = 5,450$, less 89, the solidity of the grade-prism, is 5,361.

94. WHEN THE CENTRE-HEIGHT EXCEEDS 45 FEET, after being increased as above, divide it by two, and take from the diagram quantities for the given surface-slopes. Multiply their sum by four, and subtract the grade-prism from the product. When the volume of the grade-prism for the whole length of the cut is subtracted in gross from its total solidity, as is usually the case, it is of course only necessary to record four times the quantities taken off, without any deduction.

This same method is applicable to the diagram for a uniform slope, as stated in paragraph 96. A tabular example of preliminary estimate computation is given in paragraph 119, Chapter IV.

The odometer, or measuring-wheel, furnishes, in most instances, a much better way of taking off solidities from this diagram, giving at once the total net solidity of any number of sections, whether full stations or fractional, as stated in paragraph 97; but care should be taken to draw in the horizontal line required by paragraph 97 equal to *half* the solidity of the grade-prism, since two quantities are taken off for each section.

Diagram for Sections of Uniform Slope.

95. If the surface-slope of a section be uniform, we have, by combining equations (36) and (37)—

$$S = \frac{100}{54}\frac{C^2}{(\tan R - \tan S)} + \frac{100}{54}\frac{C^2}{(\tan R + \tan S)} - \frac{50w^2}{54r}$$

$$= \frac{100}{54}\frac{C^2 2 \tan R}{(\tan^2 R - \tan^2 S)} - \frac{50w^2}{54r}.$$

Multiplying through by r, $= \cot R$, we have—

☞ $$S = \frac{100}{27}\frac{C^2}{(\tan R - \tan^2 S r)} - \frac{50w^2}{54r}. \quad (38).$$

The locus of this equation is a parabola, and the diagram constructed from it (No. 1, Plates XI. and XII.) is similar in use and appearance to No. 2 for double slopês, and is drawn on the same sheet. The value of w in the final term is taken as zero in construction.

96. Rule for Use of Diagram.—Increase all the centre-heights of the sections to be computed by the constant $\frac{w}{2r}$, as given in Table I. below. Then, *follow up the vertical line representing the centre-height to the curved line representing the surface-slope, and read off the solidity from the horizontal lines.* Centre-heights exceeding 45 feet may be taken off by the method of paragraph 94.

The solidity of the "grade-prism" has now to be subtracted for each station, and it is given in the following table for different road-beds and slopes. It is more convenient to subtract it from the total solidity of a cut, after multiplying it by the length of the cut in stations.

TABLE I.

Giving the Altitude, $\frac{w}{2r}$, and the Solidity for a length of 100 feet, of the Grade-Prism formed beneath the Road-Bed by prolonging the Side-Slopes.

SIDE-SLOPES.	1½ TO 1.		1¼ TO 1.		1 TO 1.		¼ TO 1.	
Width of Road-bed.	Alt'de.	Solidity.	Alt'de.	Solidity.	Alt'de.	Solidity.	Alt'de.	Solidity.
10 feet.	3.33	61.7	4.	74.1	5.	92.6	20.	370.4
11 "	3.67	74.7	4.4	89.6	5.5	112.	22.	448.1
12 "	4.	88.9	4.8	106.7	6.	133.3	24.	533.3
13 "	4.33	104.8	5.2	125.2	6.5	156.5	26.	625.9
14 "	4.67	121.	5.6	145.2	7.	181.4	28.	725.9
15 "	5.	138.9	6.	166.7	7.5	208.3	30.	833.3
16 "	5.33	158.	6.4	189.6	8.	237.	32.	948.1
17 "	5.67	178.4	6.8	214.1	8.5	267.6	34.	1070.4
18 "	6.	200.	7.2	240.	9.	300.	36.	1200.
19 "	6.33	222.8	7.6	267.4	9.5	334.3	38.	1337.
20 "	6.67	246.9	8.	296.3	10.	370.4	40.	1481.4
21 "	7.	272.2	8.4	326.7	10.5	408.3	42.	1633.3
22 "	7.33	298.8	8.8	358.6	11.	448.1	44.	1792.6
23 "	7.67	326.6	9.2	391.9	11.5	489.8	46.	1959.3
24 "	8.	355.5	9.6	426.7	12.	533.3	48.	2133.3
25 "	8.33	385.8	10.	463.	12.5	578.7	50.	2314.8
26 "	8.67	417.3	10.4	500.7	13.	625.9	52.	2503.7
27 "	9.	450.	10.8	540.	13.5	675.	54.	2700.
28 "	9.33	484.	11.2	580.7	14.	725.9	56.	2903.7
29 "	9.67	519.1	11.6	623.	14.5	778.7	58.	3114.8
30 "	10.	555.6	12.	666.7	15.	833.3	60.	3333.3
31 "	10.33	593.2	12.4	711.9	15.5	889.8	62.	3559.3
32 "	10.67	632.1	12.8	758.6	16.	948.1	64.	3792.6
33 "	11.	672.2	13.2	806.7	16.5	1008.3	66.	4033.3
34 "	11.33	713.6	13.6	856.3	17.	1070.4	68.	4281.4
35 "	11.67	756.2	14.	907.4	17.5	1134.3	70.	4537.
36 "	12.	800.	14.4	960.	18.	1200.	72.	4800.
37 "	12.33	845.1	14.8	1014.1	18.5	1267.6	74.	5070.4
38 "	12.67	891.4	15.2	1069.6	19.	1337.	76.	5348.1
39 "	13.	938.9	15.6	1126.7	19.5	1408.3	78.	5633.3
40 "	13.33	987.7	16.	1185.2	20.	1481.4	80.	5925.9

97. THE ODOMETER, OR MEASURING-WHEEL, is a great convenience in making up estimates from this diagram, as it enables the total net solidity of any number of stations to be taken off at once. This is a very simple instrument, and can be obtained of any dealer for about $1.50. It is simply a milled wheel, about three-quarters of an inch in diameter, turning on a screw as its axis, and with an index-point extending nearly to the periphery of the wheel to give greater definiteness in use. Starting the wheel at either end of the screw, it is run over as many vertical ordinates as desired, and then run back over a scale; evidently giving the total length of all those that have been taken, and furnishing a mechanical method of addition and subtraction. The process is very rapid and simple, and affords a great accuracy if carefully performed, rarely differing more than a small fraction of one per cent. The edge of the diagram may be used as a scale, but as the paper shrinks somewhat unequally it is better to construct one about 20,000 cubic yards long which shall average this imperfection.

When the odometer is used, the necessity of subtracting the grade-prism may be avoided, by penciling in the horizontal line equal to its solidity, and always running the odometer down to that line instead of the bottom line of the sheet.*

* This use of an odometer originated with the late John R. Gilliss, C. E., by whom 200 miles of the Union Pacific Railroad were thus estimated. See *Van Nostrand's Engineering Magazine*, vol. ii., p. 532, where

In using the odometer, start with the wheel close to the index point, and the latter on the left-hand side, and run it always *downward*, in taking off quantities, from the curve for the given surface-slope to the line representing the solidity of the grade-prism.

98. THE SOLIDITY OF HALF AND QUARTER STATIONS may be taken off in the odometer by considering that the vertical ordinate of C is four times that of $\frac{C}{2}$. Thus, if $C = 24.4$, and it is desired to take off a solid 75 feet long, run the odometer three times over the vertical line $C = 12.2$; running it each time down to the horizontal line of S equal to *one-fourth* of the solidity of the grade-prism.

Similarly, WHEN THE CENTRE-HEIGHT IS GREATER THAN THOSE SHOWN ON THE DIAGRAM, divide it by two, and run the odometer four times over the corresponding vertical line, running *three times* down to the bottom of the sheet and *once* only down to the line representing the solidity of the grade-prism.

99. SECTIONS HAVING THE CENTRE-HEIGHT ONLY GIVEN may be taken off from the Diagram for Sections of Uniform Slope; the surface-slope in that case being assumed to be level. The only advantage in so doing, over Trautwine's or other tables, is that it enables the

a description is given of an ingenious diagram for level section profile estimates, which was probably the first application of analytical geometry to such purposes.

odometer to be used, saving some time and chance for error. It is hardly necessary to suggest that in making estimates from profiles the centre-heights may be read off with the necessary constant addition already added.

100. Side-Hill Sections might be specially provided for on the same diagram as the above in a very simple way, but the error due to such sections is so very small, in making up estimates as above, that the lines would become confused. Except in very rough country such sections may be taken off as if they were thorough-cut sections, without any error of importance.

The error is, that the small triangles A and A', Fig. 24, are omitted from the solidity. A has a *maximum* solidity of from $\frac{1}{5}$ to $\frac{1}{6}$ that of the grade-prism, and A' about $\frac{1}{3}$ to $\frac{1}{4}$. Ordinarily they are much less, and the total error could hardly reach one-half of one per cent. of the whole solidity, which is counterbalanced by an error of one to three per cent. from using end-area solidities. It should be borne in mind that the importance of accuracy increases rapidly with the centre-height. An error of a tenth or two on a 20-feet centre-height, or of a half-degree in taking the surface-slope, causes a far greater error than can possibly arise from this source.

Diagram for Sections with Three or Four Surface-Slopes.

101. Fig. 25 shows such a section, the surface-slopes on each side of the section being lettered S_1 and S_2, and the horizontal distance d' being the length of the first surface-slope S_1.

The directness of the previous diagrams is unattainable for this class of sections, inasmuch as there are four variables in the equation of every half-section besides the solidity S. The process employed is to take off the solidity as if the surface-slope *nearest the centre-line* were uniform for the whole half-section, and then determine the mass a, Fig. 25, which is added or subtracted, as the case may require.

102. Through the point E draw the line EF parallel with the surface-slope S_2. Prolong also the surface-slope S_1 to H; and EHG will then be the half-section which is taken off from the Diagram for Sections of Double Slopes.

FIG. 25.

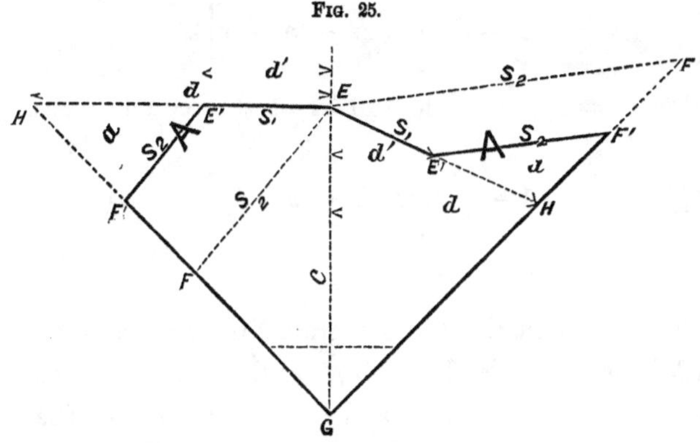

Then, to determine the mass a, the triangles EFH, or A, and $E'F'H$, or a, are similar, and it is a principle of geometry that the areas of such triangles are as the squares of any two similar sides; and, consequently, as the squares of any lines proportional to those sides. Hence, since d and $(d - d')$ are the lengths, *measured horizontally*, of the bases of the two triangles, we have—

$$A : a :: d^2 : (d - d')^2,$$

$$\frac{a}{A} = \frac{(d - d')^2}{d^2},$$

$$a = \left(\frac{d - d'}{d}\right)^2 A,$$

$$= \left(1 - \frac{d'}{d}\right)^2 A.$$

But, by equations (34) and (35), $d = \frac{C}{(\tan R \pm \tan S_1)}$. Substituting, we have—

$$a = \left[1 - \frac{d'}{C}(\tan R \pm \tan S_1)\right]^2 A.$$

Letting $C_f =$ the coefficient of A in this equation, we have—

☞ $$C_f = \left[1 - \frac{d'}{C}(\tan R \pm \tan S_1)\right]^2. \qquad (39).$$

☞ $$a = C_f A. \qquad (40).$$

103. From equation (39) Diagram No. 3 on Plates XI. and XII. is constructed, which gives the value of the coefficient for equation (40); and from equation (40) Diagram No. 4 is constructed, which gives the volume of a. In equation (39) the fraction $\frac{d'}{C}$ is considered as the variable x, and its value has to be computed before entering the diagram to take off quantities. In order to make this computation easier, *reciprocals* of $\frac{d'}{C}$, i. e., values of $\frac{C}{d'}$, are laid off on the axis of x; the value of the reciprocal being more easily obtained because d' is usually some even 10 feet, while C is in odd feet and decimals. Thus, if $C = 32.4$ and $d' = 40$, $\frac{32.4}{40}$ is much more easily obtained than $\frac{40}{32.4}$.

The volume of the mass A, which is required in equation (40), is equal to the difference of the solidities obtained by considering the half-section to have the surface-slopes S_1 and S_2 successively, as is evident from Fig. 25. In practice, this difference is best obtained by taking off the distance between the curves for those surface-slopes, on Diagram No. 2, in a pair of dividers.

104. Rule for Use of Diagrams.—The centre-height is supposed to be already increased by $\frac{w}{2r}$, the altitude of the grade-triangle, and the solidity for the half-section to have been taken off as if the inside surface-slope were uniform. Then—

First. *From Diagram No. 2, for sections of double slopes, take off in a pair of dividers, along the vertical line representing the given centre-height, the distance between the curves for the two given surface-slopes of the half-section.*

Secondly. *Divide the centre-height by the horizontal*

length of the first surface-slope, and, entering Diagram No. 3, follow up the vertical line representing the quotient, to the curved line representing the inside surface-slope; and read off a coefficient from the horizontal lines.

LASTLY. *Enter Diagram No 4, and lay off the distance taken in the dividers along the horizontal axis. Follow up the vertical line passing through the point obtained, to the inclined line representing the coefficient taken off from No. 3. The volume of the mass a is then to be read off from the horizontal lines.*

The volume of *a* is to be added to, or subtracted from, that already obtained, according as the outside slope tends to increase or diminish the area of the section. This is readily seen from an inspection of the notes. When the centre-height is greater than that shown on the diagram, divide it by two before taking off the distance in the dividers, and take four times the volume of *a* finally obtained.

Several numerical examples of this rule are given in paragraph 121, Chapter IV.

Though the above process may seem complicated, it is quite mechanical in practice, and is much more accurate than an ordinary plot, as well as more rapid and easy. Such sections are always comparatively rare, and the necessity of plotting them is an inconvenience and delay.

CHAPTER III.

APPLICATIONS TO OTHER ROAD-BEDS THAN THOSE TO WHICH THE DIAGRAMS DIRECTLY APPLY.

105. THE DIAGRAM OF CROSS-SECTIONS may be used for all road-beds which have the same side-slope as that for which it was constructed, as follows:

In Fig. 26 let w be the road-bed of the diagram

FIG. 26.

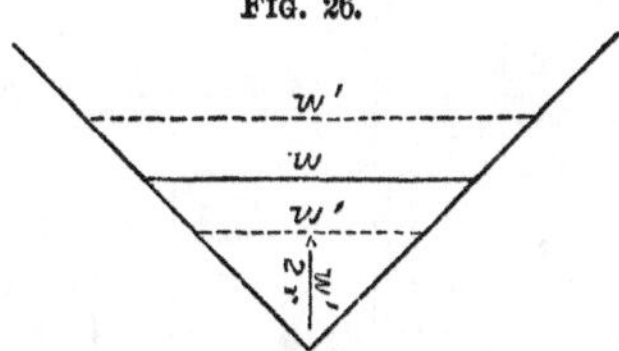

which it is desired to use for *another* road-bed, w', having the same side-slopes. It is evident from the figure that if, before entering the diagram, the vertical distance between the new road-bed and that of the plate be added to each centre-height, and the solidity due to the area inclosed between them be subtracted from the solidity obtained (or *vice versa*, according to circumstances), we shall obtain the solidity correctly. That vertical distance $= \frac{w - w'}{2r}$, and hence the area $\times \frac{100}{54} = \frac{100}{54} \frac{w - w'}{2r} \frac{w + w'}{2} = \frac{100}{54} \frac{w^2 - w'^2}{4r}$.

From these formulæ the following tables were calculated, which give the two corrections as above for all ordinary road-beds. For a fractional or other road-bed the corrections are easily deduced from the formulæ. The quantities in the table are supposed to be added *algebraically;* hence those having the *minus* sign are to be subtracted.

TABLE II.

For using Diagrams of Cross-Sections with other Road-beds.

To use the Diagram for a Road-bed of 18, 1½ to 1, with a		
Road-bed of	Add to the centre-height	And add to the solidity obtained
10 feet.	− 2.67	+ 69.1
11 "	− 2.33	+ 62.7
12 "	− 2.	+ 55.6
13 "	− 1.67	+ 47.8
14 "	− 1.33	+ 39.6
15 "	− 1.	+ 30.6
16 "	− 0.67	+ 21.
17 "	− 0.33	+ 10.8
19 "	+ 0.33	− 11.4
20 "	+ 0.67	− 23.4
21 "	+ 1.	− 36.1
22 "	+ 1.33	− 49.4
24 "	+ 2.	− 77.8
25 "	+ 2.33	− 92.9
26 "	+ 2.67	− 108.6
28 "	+ 3.33	− 142.
30 "	+ 4.	− 177.8
32 "	+ 4.67	− 216.1

To use the Diagram for a Road-bed of 14, 1½ to 1, with a		
Road-bed of	Add to the centre-height	And add to the solidity obtained
11 feet.	− 1.0	+ 23.2
17 "	+ 1.0	− 28.7
20 "	+ 2.0	− 63.
26 "	+ 4.0	− 148.2
32 "	+ 6.0	− 255.6

To use the Diagram for a Road-bed of 20, 1½ to 1, with a		
Road-bed of	Add to the centre-height	And add to the solidity obtained
11 feet.	− 3.0	+ 86.1
17 "	− 1.0	+ 34.3
23 "	+ 1.0	− 39.8
26 "	+ 2.0	− 85.2
29 "	+ 3.0	− 136.1
32 "	+ 4.0	− 192.6

TABLE II.—(*Continued.*)

To use the Diagram for a Road-bed of 18, 1½ to 1, with a			To use the Diagram for a Road-bed, 18, 1 to 1, with a		
Road-bed of	Add to the centre-height	And add to the solidity obtained	Road-bed of	Add to the centre-height	And add to the solidity obtained
16 feet.	− 0.8	+ 25.2	16 feet.	− 1.0	+ 31.4
17 "	− 0.4	+ 13.0	17 "	− 0.5	+ 16.2
19 "	+ 0.4	− 13.7	19 "	+ 0.5	− 17.1
20 "	+ 0.8	− 28.2	20 "	+ 1.0	− 35.2
22 "	+ 1.6	− 59.3	22 "	+ 2.0	− 74.1
24 "	+ 2.4	− 93.3	24 "	+ 3.0	− 116.7
25 "	+ 2.8	− 111.4	25 "	+ 3.5	− 139.4
26 "	+ 3.2	− 130.4	26 "	+ 4.0	− 163.0
28 "	+ 4.0	− 170.4	28 "	+ 5.0	− 213.0
30 "	+ 4.8	− 213.3	30 "	+ 6.0	− 266.7
32 "	+ 5.6	− 259.3	32 "	+ 7.0	− 324.1

Example.—To take off cross-sections for a road-bed of 30 feet, 1½ to 1; from the diagram for a road-bed of 18 feet, Plate III.: First add 4.0 to the centre-height of each section to be taken off. Then, after taking off the quantities, subtract 177.8 (or 178) from each, before adding them together to determine solidities; or, what is the same thing in effect, compute the solidity from the quantities as taken off, and subtract from the total volume of the cut, 177.8 multiplied by twice the entire length of the cut, from grade-point to grade-point, *along the centre-line.*

After the above constant additions have been made to each centre-height, the rules given in paragraphs 22 and 23, for taking off very large and unusual sections, and in paragraphs 24 and 25, for reducing to level section and guarding against errors in the notes, apply to the new notes without change.

106. When the side-slopes differ from those given among the diagrams, the computation is best performed from the Diagram of Triangular Prisms. The only probable side-slopes besides those given are 1¼ to 1, and 2 to 1; neither of which is likely to continue for any considerable distance. *Rock* excavation, under any circumstances, is frequently too irregular to be taken off at a single inspection, or a plate would have been added. A diagram for ¼ to 1 slopes, however, is one of very simple construction, as shown in paragraph 129.

107. Diagrams of Side-Hill Cross-Sections are applicable only to the road-beds for which they were constructed. They are given for all common single-track road-beds, and in wider road-beds for double track the grade-point is usually taken in the field, when they are readily computed from the Diagram of Triangular Prisms, by the same rule as in numerical computation.

108. The Diagram of Prismoidal Correction is general for all railway prismoids, except in the rare instances that the road-bed *differs* in the two end-sections. In all such cases the prismoid may be supposed to be divided into four triangular prismoids, in the manner indicated by the dotted lines of Fig. 16, to each of which (see par. 38) the diagram applies correctly; and we obtain as the formula of correction, letting $H =$ the sum of the side-heights—

$$C = (c - c')(D - D')\frac{l}{12} + \left(\frac{w - w'}{2}\right)(H - H')\frac{l}{12}.$$

It is evident from this equation that it is only requisite to add to the ordinary correction, the algebraic value of the last term above, as taken from the diagram. If the *slopes* only are different, and the road-bed remains the same, the correction is right, as given by the ordinary rule.

109. Diagrams for Computation by Henck's Formula may be used for other road-beds having the same side-slopes in the same way as Diagrams of Cross-Sections, and Table II. applies to them without change; but the final correction is added *only* to the quantity S_1, equation (22), taken off with the ordinary notes of each section, leaving unchanged those taken off with the dimensions touched by diagonals, viz., S_2 and S_3, equation (23). The centre-heights are increased by the same constant as that given in Table II.

When this has been done, the rules for taking off large and irregular sections given in paragraphs 72 and 73 apply to the new notes without change.

The corrections are applied only to the quantities S_1 because that is the most *convenient* way of deducting the prism between w and w', Fig. 26, the volume of which is of course the same by whatever method the solid may be computed.

110. THE DIAGRAM OF SLOPE-STAKES may be applied to all road-beds of the same side-slope as follows: Emphasize the horizontal line of d equal to half the width of the road-bed, w'; and diminish the values of the vertical lines by $\frac{w' - w}{2r}$, or, what is the same thing, increase each centre-height by the same constant before entering the diagram.

EXAMPLE.—To use the diagram for an 18-feet road-bed. Emphasize the horizontal line $d = 9$, and diminish the values of each vertical line by 2 feet, $= \frac{18 - 12}{3}$. The vertical line $c = 2$ will now be $c = 0$, and that part of the diagram to the left of it applies to side-hill sections, giving the distance out D_2, Fig. 10, when the outside slope-stake is set for an 18-feet road-bed. The diagram was constructed for a 12-feet road-bed, only as being the smallest in common use.

111. THE DIAGRAMS FOR PRELIMINARY ESTIMATES are of general application to all road-beds of $1\frac{1}{2}$ to 1, 1 to 1, and $\frac{1}{4}$ to 1 side-slopes, and may be applied indirectly to side-slopes of $1\frac{1}{4}$ to 1, $1\frac{1}{3}$ to 1, and 2 to 1, as follows: On Diagram No. 1, Plate XII., for 1 to 1 side-slopes, and uniform surface-slopes, the curve for a 24° surface-slope (or exactly 24° 5′) is equal to a *level* surface-slope with $1\frac{1}{4}$ to 1 side-slopes.* Similarly, a surface-slope of

5° $1\frac{1}{4}$ to 1, is equal to the curve for 24° 36′ on Plate XII.
10° " " " " 26° 03′ " " "
12° " " " " 26° 52′ " " "
15° " " " " 28° 18′ " " "
18° " " " " 29° 58′ " " "

These curves are easily interpolated in red ink on Plate XII.—some of them are substantially coincident with those already drawn—and it then becomes an improvised diagram for $1\frac{1}{4}$ to 1 side-slopes, to which all the rules for use that have been given apply without change. The following values will enable the curve for 20°, 22°, and 24° to be drawn on :

Centre heights ☞	8	12	16	20	24	28	32	36	40	44
Solidity, Surface-Slope, 20°	374	821	1,495	2,335	3,363	4,577	5,978	7,566	9,341	11,303
" " 22°	398	895	1,591	2,486	3,580	4,872	6,363	8,057	9,943	12,031
" " 24°	429	966	1,717	2,683	3,863	5,258	6,868	8,692	10,731	12,985

Similarly, a diagram for uniform surface-slopes and

* Calculated by the following formula :

$$\tan S = \sqrt{\frac{\tan R - \tan R'}{r} + \tan^2 S' \frac{r'}{r}},$$

in which r = the side-slopes of the diagram,
r' = the new side-slopes,
S' = the new given surface-slope,
S = the equivalent curve required in the diagram.

This formula is easily deduced from equation (38).

side-slopes of $1\frac{1}{3}$ to 1 may be improvised on Diagram No. 1 of the same plate, as follows:

A surface-slope of—

Level, $1\frac{1}{3}$ to 1,	is	equal	to the	curve for	26° 34′	on Pl. XII.
5°	"	"	"	"	27°	"
8°	"	"	"	"	27° 45′	"
10°	"	"	"	"	28° 22′	"
12°	"	"	"	"	29° 6′	"
14°	"	"	"	"	30°	"
15°	"	"	"	"	30° 26′	"
16°	"	"	"	"	30° 58′	"

Similarly, a diagram for uniform surface-slopes and side-slopes of 2 to 1 may be improvised on Diagram No. 1, of Plate XI. (for $1\frac{1}{2}$ to 1 side-slopes). A surface-slope of—

Level, 2 to 1,	is	equal	to the	curve for	18° 25′	on Pl. XI.
5°	"	"	"	"	19° 12′	"
8°	"	"	"	"	20° 20′	"
10°	"	"	"	"	21° 20′	"
12°	"	"	"	"	22° 28′	"
14°	"	"	"	"	23° 46′	"
15°	"	"	"	"	24° 26′	"
16°	"	"	"	"	25° 10′	"

When the odometer is used, the curve for a level surface-slope, in Diagram No. 1, Plate XV., may be used to make estimates for *any* unusual side-slope, by simply constructing a new scale, on which to measure the solidity taken in the odometer, which shall be to the scale of the diagram as the new side-slope, r, is to $\frac{3}{2}$. Thus, if r be $1\frac{1}{3}$ to 1, $\frac{4}{3} : \frac{3}{2} :: 8 : 9$, and $8 : 9 :: 1{,}000 : 1{,}125$. Hence, 1,000 yards on the new scale is equal to 1,125 yards by the scale of the diagram. This method, however, applies *only* to level sections.

CHAPTER IV.

OFFICE NOTES.

THOUGH no essential variations in the manner of keeping office notes are required for the use of these diagrams, a few tabular examples and hints on minor points may be of service.

112. The table on pages 110 and 111 shows a common form, perhaps as convenient as any, of keeping final estimate notes, in an ordinary transit or level book, ruled with seven columns.

In column 1 the station of each cross-section is recorded, and in column 2 the length of each solid. The field notes of the cross-sections are recorded on the second page (the wide column 8) commonly in the manner shown, although practice varies. It is also very convenient, although not absolutely necessary, to rule a narrow column (column 9) on the outside of the second page, in which to record the width from slope-stake to slope-stake, or value of *D*.

After filling in columns 1, 2, 8, and 9, the quantities in columns 3 and 4 are taken off from the proper diagrams (Plates I. and III. for the tabular example), according to the rule given in paragraph 21. Station 706 + 50, however, it will be observed, is *side-hill*, and hence is taken off from Plate VII. by the rule of paragraph 30. We obtain for it, for A and B (Fig. 9, page 27), which are in cut, 109 and 6 cubic yards = 115 cubic yards, and for C 4 cubic yards, which is in fill.

The only remaining step in determining end-area solidities is to add together the diagram quantities for each end-section of a given solid, and multiply their sum by its length, considered as the decimal part of 100 feet. Thus, in the solid from 700 + 80 to 701, we have 80, × .20 = 16; and from 701 to 701 + 20, 80 + 753 = 833, × .20 = 167. All this is identical with ordinary methods.

113. Irregular Three-Level Sections, i. e., those having more than three levels taken, are not shown in the table for the sake of simplicity, but require no essential variation in the notes. Thus, suppose the section at station 702 to have had intermediate levels taken so as to give the following notes:

$$+\frac{54.0}{30.0} \quad +\frac{30}{28.3} \quad +22.0 \quad +\frac{18}{17.3} \quad +\frac{36.0}{18.0}.$$

The section is first taken off as shown in the table, neglecting the intermediate levels, giving a solidity of 2,233, as before. The surface-lines only of the section are then plotted, which will give in this case two triangular areas, on each side of the centre, which have

TABLE OF FINAL

Road-beds, 18 *Feet in Excavation,*

(1) Station and Plus.	(2) Length of Solid.	(3) DIAGRAM QUANTITIES. Cut.	(4) DIAGRAM QUANTITIES. Fill.	(5) END-AREA SOLIDITIES. Cut.	(6) END-AREA SOLIDITIES. Fill.
700 + 80			43		(1,000 ±)
701	20	80		16	9
					(2,000 ±)
+ 20	20	753		167	
+ 50	30	1,573		697	
702	50	2,233		1,903	
+ 40	40	1,692		1,570	
703	60	762		1,472	
+ 85	85	130		758	
704	15		78	20	11
705	100		1,266	6,603	1,344
				142	
706	100		1,340	6,461	2,606
+ 25	25		62		351
		115			
+ 50	25		4	29	16
707	50		32	58	18
				87	4,374

ESTIMATE COMPUTATION.

14 *Feet in Embankment.* *Slopes,* $1\frac{1}{2}$ *to* 1.

(7)		(8)			(9)
Prismoidal Correction.		NOTES OF CROSS-SECTIONS.			Value of *D*.
		$\frac{7.0}{-0.0}$	-1.4	$\frac{11.5}{-3.0}$	18.5
	5	$\frac{15.3}{+4.2}$	$+2.0$	$\frac{9.0}{0.0}$	24.3
			8.0		33.3
82	16	$\frac{36.6}{+18.4}$	$+10.0$	$\frac{21.0}{+8.0}$	57.6
			8.0		17.7
44	13	$\frac{45.3}{+24.2}$	$+18.0$	$\frac{30.0}{+14.0}$	75.3
			4.0		14.7
18	9	$\frac{54.0}{+30.0}$	$+22.0$	$\frac{36.0}{+18.0}$	90.0
			2.6		13.8
11	4	$\frac{42.9}{+22.6}$	$+19.4$	$\frac{33.3}{+16.2}$	76.2
			6.6		26.7
54	31	$\frac{30.3}{+14.2}$	$+12.8$	$\frac{19.2}{+6.8}$	49.5
			9.6		22.5
67	57	$\frac{18.0}{+6.0}$	$+3.2$	$\frac{9.0}{+0.0}$	27.0
	7	$\frac{7.0}{-0.0}$	-2.8	$\frac{13.0}{-4.0}$	20.0
	142	$\frac{25.3}{-12.2}$	-18.0	$\frac{37.9}{-20.6}$	63.2
		$\frac{28.0}{-14.0}$	-18.6	$\frac{37.0}{-20.0}$	65.0
		$\frac{7.0}{-0.0}$	-2.0	$\frac{13.0}{-4.0}$	20.0
		$\frac{21.0}{+8.0}$	$+2.0$	$\frac{10.0}{-2.0}$	
		$\frac{7.0}{-0.0}$	-1.0	$\frac{10.6}{-2.4}$	17.6

NOTE.—The figures in small type, in columns 7, 8, and 9, are required only when the prismoidal correction is to be determined, and need not be inserted when the end-area solidity alone is sought.

been neglected in taking off the section as above. The triangle on the left side has a base 54.0 and altitude 3.0, and is to be *added* to the section, and on the right side a base 36.0 and altitude 2.7, and is to be *subtracted*. Entering the Diagram of Triangular Prisms with these notes we obtain 150 and 90 respectively, the algebraic sum of which, 60, is to be added to the 2,233 obtained above; or, what is practically better, recorded just above it, and the two bracketed together.

It is of course more convenient to plot the surfaces of all irregular sections for some considerable distances at one time, and determine for all of them at once the algebraic sum of the diagram quantities due to the intermediate levels.

Fig. 26.

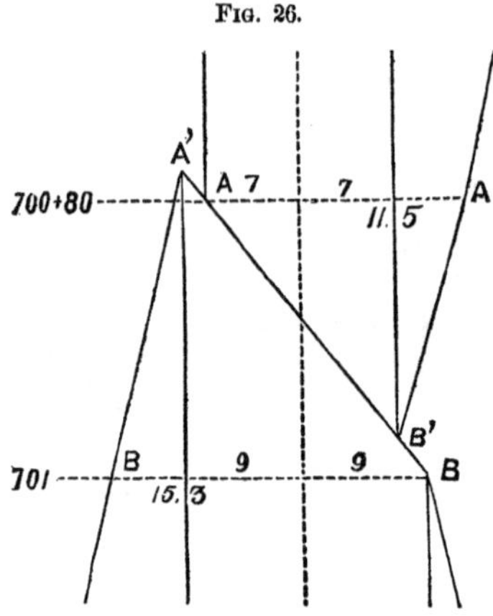

114. The cross-sections at the grade-passages from cut to fill are not taken with strict correctness in the table above, as is evident from Fig. 26, which gives a plan of the sections at 700 + 80 and 701. When the road-beds in excavation and embankment are of unequal width, there must always be some *overlapping* of the pyramids AAB' and BBA' on sloping ground; so that, besides taking the sections AA and BB, the grade-points A' and B' should be also determined, but such precision is of little or no importance with ordinary road-beds. It is more usual in practice, however, to omit the section AA at 700 + 80, and take instead of it a section at the grade-point A' where the cut runs out. In that case, instead of the field notes recorded at 700 + 80, we should have had, perhaps—

$$700 + 78 \quad \Big\| \quad + \frac{9.0}{0.0} \qquad - 1.6 \qquad - \frac{12.4}{3.6}$$

This section, as now recorded, is *side-hill*, since the left slope-stake is set for the cut road-bed, but, as the embankment is not to be carried out to that width, it is necessary to substitute for 9.0 the distance out to a fill slope-stake, if set. This may be estimated by inspection at 7.6, giving for the total width of the cross-section 20.0, instead of 21.4.

This practice gives the volume of the excavation more correctly than that shown in the notes, and is in every way better. The latter was only used for simplicity.

115. The Prismoidal Correction may be determined at any time when convenient, or omitted altogether when deemed unimportant. The small figures in columns 8 and 9 are the required differences in the given dimensions, and entering the Diagram of Prismoidal Correction (Plate IX.) with those differences, the quantities in small figures in column 7 are obtained. These quantities, after being multiplied by the decimal length of each solid regarded as the fraction of a station, give the required corrections; the sum total of which, for each cut or for a whole section, is to be subtracted in gross from the end-area solidity. No corrections are shown for the fill, as embankment is not commonly estimated.

In the pyramidal solids at the ends of the cuts the correction is given by simply taking one-third of the end-area solidity. This is a general rule for all pyramids, and hence applies also to the solids due to the side-hill section at 706 + 50, the corrections for which are not shown in the table.

By the "Method of Centre-Heights" the centre-heights only need be subtracted from each other, and the corrections may then either be read off from the curve on the diagram, or, somewhat more easily, taken from the table in Appendix B. Column 7 is then made out as before. The correction obtained for the cut given is 148 cubic yards, an error of 6 cubic yards, tending to deficiency; but were the position of the pluses varied a little, the error would be as much the other way.*

* The above tabular example furnishes another illustration of the relative accuracy of the "Method of Centre-Heights," and of "Equiva-

116. When the road-bed of the earthwork to be computed is different from those for which plates are given, the only change required in the notes is to add a constant quantity taken from Table II., page 100, to the centre-height of each section. Thus, in the above tabular example, if the road-bed in excavation had been 24 feet instead of 18, each centre-height would require to be increased by 2.0. The diagram quantities are then taken off and recorded, *in red ink*, in column 3, and the constant quantity 77.8 (or 78), as given in Table II., subtracted from each one before proceeding to fill up column 5. Or, column 5 may be filled up without subtracting the constant 78 from each, and the sum total of each cut diminished by 77.8 multiplied into twice the length of the cut, from grade-point to grade-point, *along the centre line*. This will evidently give an identical result. In the cut given above, this length is (703 + 92.5) — (700 + 90) = 3.025 stations.

The prismoidal correction is independent of the width of road-bed.

lent Level Sections." By the former method, instead of the true quantities shown in column 7 above, viz., 82, 44, 18, 11, 54, 67, we obtain, from the Diagram of Prismoidal Correction, the following: 59, 59, 15, 6, 40, 85; showing that the solidity obtained by that method will be in error on each solid respectively as follows, viz., + 23, — 15, + 3, + 5, + 14, — 18, or a total error of 12 cubic yards in excess. By the "Method of Level Sections" the errors will be successively, — 1, — 2, — 0, — 1, — 2, — 3, or 9 cubic yards in deficiency. The tendency of the cumulative error in the latter method to overbalance the much greater errors in the former is evident.

117. The table on page 117 illustrates the method of keeping notes for Computation by Henck's Formula, and is the same cut as that given in Henck's "Field-book for Engineers," page 105. The road-bed of the cut being 28 feet, the constant quantity 3.33 is first added to each centre-height, as given in Table II., page 100, and the computation may then be made from Plate X. After taking off the quantities, the quantities S_1 are diminished by 142, also given in Table II.; the original quantities being shown in small figures beside the corrected ones. Instead of correcting them individually, the product of 142 into twice the length of the cut might have been subtracted from the total volume as made up from the original quantities with the same result.

The comparative results from Henck's and the prismoidal formulæ may be illustrated by the cut in the tabular example of paragraph 112. The table on page 118 shows these results on each solid, and on the sum total. The diagonals may evidently be arranged in four different positions, which are indicated by the headings to the columns of solidity.

It will be observed that the mean of the solidities obtained by taking the diagonals in any given direction and then reversing them is precisely identical with that given by the prismoidal formula, barring slight errors of observation. It is easily demonstrated that this should be so by comparing the formulæ. The differences in any event are very small except in extraordi-

TABLE OF FINAL ESTIMATE COMPUTATION BY HENCK'S FORMULA.

Road-bed 28 feet, Slopes $1\frac{1}{2}$ to 1.

FIELD-NOTES, AS GIVEN ON PAGE 105, HENCK'S "FIELD-BOOK."							SOLIDITIES IN CUBIC YARDS FROM DIAGRAM.			
Station.	h	d	c	d'	h'	$(d + d')$	S_1	S_2	S_3	Volume of each solid
0	2.0	17.0	5,33 2.0	17.0	2.0	34.0	188 46			
1	3.0	18.5	7,33 4.0	21.5	5.0	40.0	280 138	90	100	374
2	4.0	20.0	8,33 5.0	23.0	6.0	43.0	330 188	115	139	580
3	6.0	23.0	9,33 6.0	26.0	8.0	49.0	434 272	153	180	793
4	5.0	21.5	9,33 6.0	24.5	7.0	46.0	386 244	153	181	850
5	4.0	20.0	9,33 6.0	20.0	4.0	40.0	329 187	140	181	752
6	1.0	15.5	7,33 4.0	18.5	3.0	34.0	230 88	97	115	487
Total Solidity = 1,163 − (46 + 88)							+ 1,163	+ 748	+ 896	=3,836 c. y.
Total Solidity as given in the volume,							103,533 cubic ft., equal to 3834.56 c. y.			

Table of Comparison.

STATION.	Solidity by Prismoidal Formula, as given in par. 112.	Solidity by Henck's Formula.					
		Direction of Diagonals.		MEAN.	Direction of Diagonals.		MEAN.
		∧	∨		//	//	
700 + 80							
701	11	11	11	11.	11	11	11.
+ 20	151	155	146	150.5	149	152	151.5
+ 50	684	669	701	685.	697	673	684.
702	1,894	1,902	1,887	1894.5	1,893	1,895	1894.
+ 40	1,566	1,547	1,585	1566.	1,589	1,542	1565.5
703	1,441	1,407	1,473	1440.	1,422	1,459	1440.5
+ 85	701	714	686	700.	682	719	700.5
704	13	13	13	13.	13	13	13.
Totals,	6,461	6,418	6,502	6460.	6,456	6,464	6460.

nary solids; and on such three levels to a cross-section are rarely sufficient.

118. THE OFFICE NOTES OF IRREGULAR EARTHWORK, such as sidings, borrow-pits, etc., etc., are usually kept in a separate book, in which a little table is made out, for each section or solid, of the dimensions which require to be multiplied together. This process is the best also for computation by diagrams; those dimensions being recorded with which it is necessary to enter the diagram, as in the numerical example given below, which gives the process of computation, both numerically and by diagrams, for the sections shown in Fig. 19. It will be seen that there is no essential difference between them except in the simplicity and conciseness of the latter process.

Table for computing the Section shown in Fig. 19.

As arranged For Numerical Computation.		As arranged For Computation by Diagrams.		
Dimensions to be multiplied together.	Computed Areas in square feet.	Dimensions to enter the Diagram with.		Quantities read off, c. y.
$\frac{25.0 \times 17.0}{2} =$	212.5	25.0	17.0	393
$\frac{20.0 \times 16.4}{2} =$	164.	20.0	16.4	304
$\frac{25.0 \times 12.3}{2} =$	153.75	25.0	12.3	286
$\frac{30.0 \times 10.6}{2} =$	159.	30.0	10.6	295
$\frac{25.0 \times 10.9}{2} =$	136.25	25.0	10.9	252
$\frac{32.5 \times 10.1}{2} =$	164.125	32.5	10.1	304
Area in square feet.	989.625	Solidity for ½ Station		1,834

The computed area above, added to a similar area for the next station, must be multiplied by 100 and divided by 2 and by 27 to obtain the solidity. The sum of the diagram quantities has only to be added to a similar sum for the next station to give the same result at once.

In the above table the section is subdivided in the manner shown by the dotted lines of Fig. 19. It of course might equally well be subdivided into trapezoids, or in any other manner which would give the area correctly. The tabular forms for irregular earthwork of any class differ only in detail from the above, and it is of course advisable to make them up at once for the entire estimate. Their greater or less number is then a very small item.

119. As an example of Preliminary Estimate Computation, a very irregular cut is given in the table below. Where but one surface-slope is given at a station, it is supposed to be uniform for the whole section.

TABLE OF PRELIMINARY ESTIMATE COMPUTATION.

Road-bed, 18 ; *Slopes*, 1½ *to* 1.

FIELD NOTES.				COMPUTATION FROM PLATE XI.			
Station	Slopes *L*.	Centre H'g'ts.	Slopes *R*.	Diagram Quanti's.	Net Solidities.	Corrections. +	Corrections. —
0 + 50		6.0 ~~0~~	10°	230	115		
1		8.0 ~~2.0~~	15°	440	330		
2	20 – 5°, – 10°	23.0 ~~17.0~~	+ 20°	1,160 3,240	4,400	40	
3	– 16°	42.0 ~~36.0~~	30 + 20°, + 15°	3,430 10,780	10,656		1180 ~~1570~~
3 + 50	– 14°	26.0 ~~20.0~~	+ 18°	1,360 3,680	2,520		
4	5° –10°, – 15°	46.0 ~~40.0~~	40 + 14°, + 10°	4,200 9,360	10,170	+ 0	420 ~~560~~
5	– 10°	63.0 ~~57.0~~	+ 17°	8,720 20,360	29,080		
6	– 8°	36.0 ~~30.0~~	+ 14°	2,970 5,740	8,710		
7		7.0 ~~1.0~~	15°	340	340		
Totals,					66,321	+ 170	–1600
Less volume of grade-prism = 200 × 6.5,					1,300		
Total, exclusive of exterior surface-slopes, *Corrections*, + 170 – 1,600, =					65,021 –1,430		
Final solidity,					63,591	cubic yards.	

To make the computation, all the centre-heights are first increased by 6, equal to the altitude of the grade-prism, or $\frac{w}{2r}$, as taken from Table I., page 91. The solidities recorded in the first column are then taken off from Plate X. (neglecting entirely the exterior surface-slopes where more than one is given on each side of the centre), and the sum of the diagram quantities for each section transferred to the second column; except when there is a *plus* station on either side of the section, when the length of the solid due to the section is taken as equal to half that between the two sections on either side, and the corresponding fractional part of the solidity taken. Thus, section 3 is $\frac{3}{4}$ of a station long, and $3 + 50$ is half a station long. This gives the same (end-area) solidity as would be obtained by the ordinary process of averaging, but is a little more convenient when "pluses" are comparatively rare.

The end-sections are assumed to run out to zero at the next section, giving a slight tendency to excess which counterbalances the error due to taking off side-hill sections as if thorough-cut. Thus, section $0 + 50$ and 7 are side-hill, and 30 or 40 cubic yards at each are omitted from the calculation.

The volume of the grade-prism, 200 cubic yards, is given by Table I., page 91. The second column in the table above may often be omitted in practice, when fractional stations are rare.

120. By the odometer the total net solidity of 65,021 cubic yards is determined at once within a small fraction. The cut in question was thus taken off three times independently with the following results: 64,700, 64,650, and 64,800 cubic yards; which affords an indication of the probable error, the major portion of which is evidently due to unequal contraction of paper. This is an unavoidable source of error, but hardly important.

The above cut is exceptionally difficult to take off in an odometer, owing to its irregular nature; but where most of the sections have a uniform slope, and are a full station long, the process is very rapid. In the above cut, sections 3 and 4 are only 75 feet long. Hence (par. 98) the centre-height has to be divided by two, and the vertical ordinates taken three times on that new centre-height, running the odometer each time down to a point equal to one-fourth the solidity of the grade-prism. Similarly, 3 + 50 is only 50 feet long, and the vertical ordinate of half the centre-height is taken *twice*. The centre-height of section 5 has to be divided by 2, and. its vertical ordinate taken four times, because it exceeds the limits of the diagram.

121. The corrections for outside surface-slopes, in the last two columns, seldom require a separate column in practice. The following is the process of computing them, beginning at section 2:

Take, on Diagram No. 2, Plate XI., for sections of double slopes, the vertical line for the given centre-height, 23.0, and take off in dividers the distance along it between the curves for $+ 5°$ and $- 10°$. Also, entering Diagram No. 3, divide the centre-height, 23.0, by 20, giving 1.15, and follow up the vertical line 1.15 to the curved line $- 10°$, when the coefficient obtained is .07. Then, *finally*, enter No. 4, and laying off the distance in the dividers on the horizontal axis, the vertical line through the point obtained is 540. Follow it up to the inclined line for the coefficient .07, when the correction for a full station due to the mass *a*, Fig. 25, is read off as 40 cubic yards, which is in this case to be *added* to the section.

For station 3, take off in dividers on No. 2 the distance between the curves for $+ 20°$ and $+ 15°$. Then, $\frac{42.0}{30} = 1.4$, and the coefficient obtained from No. 3 is .61. Entering No. 4 with that coefficient and the distance in the dividers, the correction obtained is 1,570 yards, which is to be subtracted in this instance. On the right side of section 4, $\frac{46}{40} = 1.15$, and the coefficient obtained is .41; the distance in the dividers is 340 cubic yards (*half* the centre-height or 23.0 having been taken, as 46.0 is not shown in the plate), and the correction obtained from Diagram No. 4 is 140 yards, which is to be multiplied by 4, giving 420, and subtracted, in this instance.

On the left side of section 4, $\frac{46.0}{50} = .92$, and entering No. 3 with .92 and $-15°$, the coefficient is found to be less than zero. This indicates that the inside slope extends beyond the limits of the cross-section, and consequently no correction is required.

When, as in sections 3 and 4, the corrections apply to sections which are not a full station long, the proper fractional part, in the above cases $\frac{3}{4}$, is to be taken.

A somewhat complicated example has been purposely chosen, but there are very few sections in ordinary heavy work for which two slopes are not sufficient. There is, however, always a considerable error involved in assuming the surface-slope to be *uniform*, when the slopes on each side of the centre vary as little even as two or three degrees. It is impossible to *average* the slopes so as to materially reduce this error, although that may easily be done when there are two or more slopes on one side of the centre.

CHAPTER V.

CONSTRUCTION OF DIAGRAMS.

122. THE construction of the diagrams on the plates, or others similar to them, is a very simple matter, requiring hardly any computation in most instances, and little attention to their theoretical basis. The utmost attention to accuracy is needed, however, and a little ingenuity in avoiding unnecessary labor.

The following remarks are of general application to all such construction. Special peculiarities of the different diagrams are considered individually hereafter.

123. The method invariably employed in constructing diagrams of this kind is to lay off points for the inclined lines or curves (i. e., points for the locus of the equation for different values of the constant, or *third variable*) on a certain number of vertical lines. When the equation is of the first degree (that is to say, including also the constant or *third* variable), these points —on any one vertical line—are always at equal distances apart; and, if that line be *properly selected*, they will also be at *even* (or non-fractional) distances apart, so that they can then be laid off almost mechanically.

124. Thus, the equation of the DIAGRAM OF TRIANGULAR PRISMS is—

$$S = \frac{50c}{54} D.$$

If we let $D = 54$, we have $S = 50c$. Consequently, on the vertical line 54, points for $c = 1, 2, 3$, etc., may be obtained without calculation by simply taking its intersection with every fifth horizontal line successively. If we let $D = \frac{8}{10}$ of $54 = 43.2$, we have $S = 40c$, and every *fourth* horizontal line gives a point for the lines of c; and, similarly, on the vertical lines $D = 10.8$; 16.2 ; 21.6 ; 27, etc. (which are all multiples of 5.4), the points for lines of c varying by one foot will be 10, 15, 20, 25, etc., cubic yards apart. The fine *intermediate* lines for variations of two-tenths are of course one-fifth of this distance apart in all cases.

So, also, in constructing the DIAGRAM FOR CORRECTION ON CURVES, from equation (29), page 77; letting $D = 17.19$, we have $C_D = \frac{d - d'}{100}$, and the points on the corresponding vertical, for values of $(d - d')$ varying by 10 feet, will be .1 by scale apart; and on the vertical lines $D = 34.38$; 51.57, etc., they will be .2; .3, etc. For the same reason, in constructing the Diagram of Prismoidal Correction, those verticals are selected which are multiples of 3.24.

125. In this manner, a diagram to solve any equation of the form $y = azx$ can be constructed without any computation whatever—other than mental—by simply selecting those vertical lines of x which are multiples of the reciprocal of a; when if z be increased successively by one, the corresponding increase, ax, in the value of the equation (which may be termed the *constant difference*) is always integral.

This is also to a great extent the case with *all* equations of the first degree. The same process may be always used after the determination of points on a single line. Thus, the most complicated equation of the first degree between three variables must have the form $y = a(z + b)x + c$. If z be increased successively by one the constant difference is ax, the same as in the equation above; and if x be also a multiple of the reciprocal of a, the value of ax will be integral.

126. In theory, two points are sufficient for each line, but for accurate work points should not be more than two to four inches apart, as ordinary cross-section paper is apt to be a little warped. On the Diagram of Cross-Sections and of Triangular Prisms points are laid off 2.7 inches apart horizontally, on the Diagram of Prismoidal Correction 3.24 inches, and on the Diagram for Computation by Henck's Formula 4.05 inches, and so on, according as the equation requires. Always construct first the lines for *even* values of the constant or third variable, and afterward fill in points for and draw in the intermediate lines. The latter process is almost always quite mechanical. Thus, when the lines for even feet are 40 yards apart, the lines for two-tenths will be 8 yards apart, and each successive point will be 8, 6, 4, 2, 0 cubic yards above the fine horizontal lines; if the intermediate lines are 7 yards apart, each point will be 7, 4, 1, 8, 5, etc., successively, and so on, *ad infinitum.*

127. Spacing-dividers should *not* be used; they lead to cumulative errors. Each point should be laid off independently; and a good draughtsman can do this rapidly, and also, for even quantities, with great accuracy. Points should always be laid off near the ends of the lines, even if fractional values are required, or it is necessary to pencil on a vertical line from two to five-tenths beyond the edge of the sheet, as the paper is usually most warped near the edges. The straight-edge will rarely lie *exactly* on the points of a straight line, and the drawing-pen should be slightly swayed to pass through every point as nearly as may be. The utmost care is requisite, but slight errors will balance each other, and when equal to a change of two or three hundreds in the notes will be very conspicuous. None of the lines should be penciled in, as it leads to inaccuracy, and it is best to draw only half of a long line at once. Much of the cross-section paper in the market is quite unfit for this purpose. Queen & Co.'s "Plate G," 16 × 20 inches, is the most suitable, and all the plates here given were constructed on it, being afterward reduced to $14 \times 17\frac{1}{2}$ inches by photo-lithography.

128. IN CONSTRUCTING A DIAGRAM OF CROSS-SECTIONS, make out a table similar to the table below, which gives points for the bottom line of each subdivision, from which to begin laying off points *without* calculation. The equation of the diagram is—

$$S = \frac{50}{54}\left(c + \frac{w}{2r}\right) D - \frac{25w^2}{54r},$$

and points are hence most conveniently determined on lines of D which are multiples of 5.4; the vertical distance between lines of c varying by one foot being then $5\frac{D}{5.4}$; and for lines varying by 0.2, $\frac{D}{5.4}$. Fill in thus the first three columns of the table to the highest values required, and head the succeeding columns with the values of c with which it is designed to begin each subdivision of the diagram. These values of c should always be about the same for all road-beds of the same side-slopes, and may be judged from the plates.

To fill up the vertical columns of c, any *one* quantity in them will $= 5\frac{D}{5.4}\left(c + \frac{w}{2r}\right)$ minus the final constant, equal to half the solidity of the grade-prism in Table I., page 91. Each *succeeding* quantity in the same column, since they are for values of D increasing by 5.4, will then be given by adding to it $5\left(c + \frac{w}{2r}\right)$. Head each column of c with this constant difference, and the value of any one quantity in it will then also be given by multiplying the constant difference by the number in column 3 $\left(\text{which is } \frac{D}{5.4}\right)$, and subtracting half the volume of the grade-prism given in Table I. Compute in this way the first and last quantity required in each column, and fill up the table by addition.

To find what quantities are required, compute the width of a level section of the centre-height at the head of each column (which is equal to $2cr + w$), and let the smallest value of D in each subdivision of the diagram be two or three feet less than this, the greatest of course being 32 feet greater. Then, each line of c in the table will be the top line of one subdivision and the bottom line of the next above, and the lines of D, on which points are required, are easily determined.

Having now, after completing the table, the notes for the top and bottom line of each subdivision, these values are plotted, and the remaining points filled in with the constant differences of columns 2 and 3, as already described in paragraph 126.

The Curve of Level Section is constructed, after completing the rest of the diagram, by simply passing it through the proper intersection-points of c and D, beginning at $c = 0$, $D = w$, in the lower left-hand corner.

New values are assigned during construction to the horizontal lines in each subdivision, as found convenient.

TABLE OF NOTES

For constructing a Diagram of Cross-Sections.
Road-bed 24, *Slopes* $1\frac{1}{2}$ *to* 1.

Lines of *D*, on which points are laid off.	Vertical distance apart of lines of *c* varying by		COLUMNS OF *c*. (Figures in parentheses, constant differences between quantities in each column of *c*.)					
			(40)	(75)	(105)	(130)	(135)	(175)
	1.0	0.2	*c* = 0	*c* = 7	*c* = 13	*c* = 18	*c* = 23	*c* = 27
27.0	25	5	22	197				
32.4	30	6	62	272				
37.8	35	7	102	347				
43.2	40	8	142	422	662			
48.6	45	9	182	497	767			
54.0	50	10	222	572	872			
59.4	55	11		647	977	1,252		
64.8	60	12		722	1,082	1,382		
70.2	65	13		797	1,187	1,512		
75.6	70	14			1,292	1,642	1,992	
81.0	75	15			1,397	1,772	2,147	
86.4	80	16			1,502	1,902	2,302	
91.8	85	17			1,607	2,032	2,457	2,797
97.2	90	18				2,162	2,612	2,972
102.6	95	19				2,292	7,767	3,147
108.0	100	20				2,422	2,922	3,322
113.4	105	21					3,077	3,497
118.8	110	22					3,232	3,672
124.2	115	23						
129.6	120	24						
135.0	125	25						
Width on top of *level sec'ns.*			24.	45.	63.	78.	93.	105.
Values of *D* in each subdivision.			(24 to 56)	(42 to 74)	(60 to 92)	(76 to 108)	(91 to 123)	(123 to

(Solidity of "grade-prism," 355.6.)

129. In Rock Excavation, since the side-slopes are very steep, a diagram is much smaller and more easily constructed than for flatter slopes—owing to the fact that the values of *D* are much reduced and less liable to wide variations. With $\frac{1}{4}$ to 1 side-slopes, the width of a section of 40 feet centre-height is only 20 feet greater than the width of road-bed, and a diagram for that ratio of slope may be constructed on a sheet of cross-section paper about 12 × 14 inches. The following table gives the requisite notes for constructing such a diagram, for a road-bed of 18 feet. The first subdivision extends to 20 feet centre-heights, and the next to 40 feet; and, after completing the diagram to 40 feet, there will be room in the upper left-hand corner to extend it to 50 or more feet if desired.

TABLE OF NOTES

For constructing a Diagram of Cross-Sections.

Road-bed 18, *Slopes* $\frac{1}{4}$ *to* 1.

Lines of D, on which points are laid off.	Vertical distance apart of lines of c varying by		COLUMNS OF C. (Figures in parentheses, constant differences between quantities in each column of c.)		
			(90)	(140)	(190)
	1.0	0.2	$c = 0$	$c = 20$	$c = 40$
18.9	17.5	3.5	30	380	
21.6	20.	4.	120	520	
24.3	22.5	4.5	210	660	
27.	25.	5.	300	800	1,300
29.7	27.5	5.5	390	940	1,490
32.4	30.	6.		1,080	1,680
35.1	32.5	6.5		1,220	1,870
37.8	35.	7.		1,360	2,060
Width on top of *level sections.*			18.	28.	38.
Values of D in each subdivision.			(18—30) (27—39)		

(Solidity of "grade-prism," 1,200.)

Owing to the greater value of the term $\frac{w}{2r}$ in the equation of this diagram, the lines of c would be at a much greater inclination than on one for flatter side-slopes, and to obviate this practical inconvenience the horizontal scale of the diagram should be doubled. This in turn requires the number of vertical lines on which points are laid off to be doubled (in order to retain the same interval in inches between them), and, on this account, the vertical distance apart of the points for the inclined lines is not always in even yards, as will be seen in the table; but, with a little ingenuity, the process of laying off points is about equally mechanical.

A diagram for ¼ to 1 slopes is not given among the plates, partly because it is so easily constructed, and partly because rock excavation is so frequently irregular, and always easily computed from the Diagram of Triangular Prisms.

130. The Construction of the Diagram for Computation by Henck's Formula differs only in detail from the preceding. In computation by that method, quantities are to be taken off, not only with the centre-height and *total* width on top, but also with the centre-height and one of the distances out. Hence a much greater "range" is required, and it will be observed, on Plate X., that centre-heights of 30 feet are shown even in the first subdivision, giving a somewhat peculiar appearance to the diagram.

The equations which it is required to solve (equations (22) and (23), page 66) have the form—

$$S = \frac{50}{81}\left(c + \frac{w}{2r}\right)D - \frac{25w^2}{108r},$$

the final constant being equal to $\frac{1}{4}$ the solidity of the grade-prism as given in Table I. Points should evidently be determined on lines of D which are multiples of 8.1, the vertical distance between lines of c varying by 0.2, being then $\frac{D}{8.1} = 1, 2, 3,$ etc., and for lines varying by one foot, $\frac{D}{8.1} \times 5 = 5, 10, 15,$ etc. Columns 1, 2, and 3 of the table below are thus filled up, and are always the same for all road-beds. Then, for a road-bed of 24, $1\frac{1}{2}$ to 1, substituting for $\frac{w}{2r}$ its value, 8, and for the final constant its value, 88.9 (or 89), we have—

$$S = \frac{50}{81}(c + 8)\,D - 89.$$

The lines of c which will be most convenient for construction are $c = 0, 10, 20, 25, 30$, and remaining columns of the table below are so headed. Any *one* quantity in them is equal to $5\,\frac{D}{8.1}(c + 8) - 89$, and each *succeeding* quantity in the same column will be given by adding to it $5\,(c + 8)$. Head each column of c with this constant difference, and any one quantity in it can then also be obtained by multiplying the number in column 3, $\left(\text{which is } \frac{D}{8.1}\right)$ by the constant difference, and subtracting the constant 89. Compute in this way the first and last quantity of each column, and fill up the table by addition.

All this process, together with the subsequent steps of construction, is the same as in the Diagram of Cross-Sections, except that in this case the upper line of the first subdivision is the irregular line shown in Fig. 26, instead of a given line of c, and that the upper line of the next subdivision is $c = 30$. Draw on the broken subdivision line, in the same position exactly as in Plate X. (for any road-bed), before constructing the lines, which should then be extended to it in both subdivisions, leaving only space in the upper one to write in the values of the vertical lines. The points laid off for each subdivision should be extended a little above and below this division line, to enable the inclined lines to be more accurately constructed, and the same remark applies to points for the two upper subdivisions.

The horizontal distance in this diagram between the verticals on which points are laid off is the widest allowable, and, near the edges of the sheet, it is well to fill in points on lines half-way between those shown in the table below. Thus, the line $D = 36.45$ may be interpolated, and the vertical distance apart on it of lines of c varying by 1.0 will be 22.5, and by 0.2, 4.5. The values to be interpolated in the remaining columns are equal to the half-sum of those above and below it.

TABLE OF NOTES

For Construction of a Diagram for Computation by Henck's Formula. Road-bed 24, Slopes $1\frac{1}{2}$ *to* 1.

Vertical lines of D, on which points are laid off.	Vertical distance apart of lines of c varying by		COLUMNS OF c. (Figures in parentheses, constant differences between quantities in each column of c.)				
			(40)	(90)	(140)	(165)	(190)
	1.0	0.2	$c = 0.$	$c = 10$	$c = 20$	$c = 25$	$c = 30$
8.1	5	1	−49	1	51	76	101
16.2	10	2	− 9	91	191	241	291
24.3	15	3	31	181	331	406	481
32.4	20	4	71	271	471	571	671
40.5	25	5	111	361	611	736	861
48.6	30	6	151	451	751	901	1,051
56.7	35	7	191	541	891	1,066	1,241
64.8	40	8	231	631	1,031	1,231	1,431
72.9	45	9	271	721	1,171	1,396	1,621
81.0	50	10	311	811	1,311	1,561	1,811
89.1	55	11	351	901	1,451	1,726	2,001
97.2	60	12	391	991	1,591	1,891	2,191
105.3	65	13	431	1,081	1,731	2,056	2,381
113.4	70	14	471	1,171	1,871	2,221	2,571
121.5	75	15	511	1,261	2,011	2,386	2,761
129.6	80	16	551	1,351	2,151	2,551	2,951

(Solidity of "grade-prism," 355.6.)

131. In constructing a Diagram for Side-Hill Cross-Sections, the sub-diagrams B and C consist of curved lines; and values of the equations of area are computed for a range from 0 to 6 feet of c and h_2, multiplied by $\frac{100}{54}$ with any of the tables in common use, and plotted on the vertical lines of D_2 corresponding to the given value of h_2. The computations are few and short, if made in tabular form, but other diagrams than those given will hardly be required, as all such sections may be taken off from the Diagram of Triangular Prisms.

The diagram of A is identical in principle with the Diagram of Cross-Sections. It is only necessary to determine from the equation two or three points for the line $c = 0$ on vertical lines of D which are multiples of 5.4. The remaining points are filled in mechanically, as stated in paragraphs 124 and 126.

132. THE DIAGRAM OF SLOPE-STAKES is constructed with values obtained from a table made out as below, for each degree of surface-slope, plus or minus, required:

Side-Slopes $1\frac{1}{4}$ *to* 1. (*tan* R = 8.)

1	Surf'ce-Slope S	+ 20°	+ 10°	0°	− 10°	− 20°
2	$\tan S$	.364	.1763	.0	.1763	.364
3	$\tan R \pm \tan S$	.436	.6237	.8	.9763	1.164
4	$\frac{1}{\tan R \pm \tan S}$	2.294	1.604	1.25	1.024	0.86
5	$c + \frac{w}{2r} =$ 5	11.47	8.02	6.25	5.12	4.30
	" 10	22.94	16.04	12.5	10.24	8.60
	" 15	34.41	24.06	18.75	15.36	12.90
	" 20	45.88	32.08	25.	20.48	17.20
	" 50	114.70	80.20	62.5	51.20	43.00

Line 4 is the reciprocal of line 3, and is taken off from a table of reciprocals; the next line is ten halves of it, and the remaining lines are given by constant addition. The whole computation can be made in an hour, though more extensive than is required for the previous diagrams.

In constructing for any given road-bed, the term $\frac{w}{2r}$, equations (34) and (35), is subtracted from 5, 10, 15, etc., to give the proper vertical lines of c on which to lay off the values above; the verticals for *minus* values of c applying to *side-hill* sections, and giving the point e, Fig. 23. The line $d = \frac{w}{2}$ is the lower limit of the diagram in practice, because, if the inclined lines were extended below it, they would give the point e'. Hence, when an intersection cannot be found above that line, it *indicates* that the section is side-hill, and the distance out is given by the diagram for the other road-bed.

133. Construction of the Diagrams for Preliminary Estimates.—Nos. 1 and 2 are composed of parabolas, and constructed by computing points on each curve at intervals of four feet. About half those required are computed by logarithms, and the remainder obtained by taking one-fourth of the computed values. Any geometrical construction of the curves, if carefully done, would require more labor than computation.

The Diagram of Coefficients, No. 3, is constructed from a little table, made as follows: Column 1 gives values of $\frac{d'}{C}$, varying by .1 up to about 1.5; column 2, *reciprocals* of column 1, giving the vertical lines on which the tabular values are plotted; and the remaining columns are headed with plus and minus surface-slopes varying by 5°, and with the corresponding value of (tan $R \pm$ tan S). The table is then filled up by addition, and squared with a table of squares. Three decimals are ample.

Diagram No. 4 is of the simplest possible description; both the variables, A and a, being the same quantities, and laid off to the same scale. The highest value of C_f is 1, and the locus of the equation is then at an angle of 45° with the axes. Points for other values of C_f varying by .1 are laid off on almost any one of the verticals by inspection.

APPENDIX A.

COMPARISON OF DIFFERENT METHODS OF APPLYING THE PRISMOIDAL FORMULA TO EARTHWORK COMPUTATIONS.

THIS Appendix is added to compare the different methods of applying the prismoidal formula, first, to the computation of three-level sections; and, secondly, to all forms of irregular sections in road-bed excavation; that is to say, earthwork with regular side-slopes.

The methods which may be used in computing three-level earthwork are—

1. *The Method of par.* 37—based on a direct determination of the dimensions of the mid-section, and the standard for all.

2. *The Method of Level Sections;* defined in paragraphs 35 and 39.

3. *The Method of Centre-Heights;* defined in par. 41.

We will first compare the Method of Level Sections with a direct determination of the dimensions of the mid-section.

Taking any "three-level" prismoid, suppose the grade-prism, so called, formed by prolonging the side-slopes to an intersection below the road-bed, to be included in the solid. This is done partly to simplify the demonstration, and partly because it has been claimed, by a recent writer,* that its introduction rectifies the error in the Method of Level Sections. It will be seen, however, that this is incorrect, and that its introduction or exclusion has no effect whatever on the comparative result.

Letting A and A' = the areas of the two end-sections; c_L and c'_L = the centre-height of *level* sections of equivalent area $\left(\text{always equal to } \sqrt{\frac{A}{r}}\right)$, and D and c = the width on top and the centre-heights of the end-sections, we have—

$$A = \frac{Dc}{2}, \qquad A' = \frac{D'c'}{2},$$

$$c_L = \sqrt{\frac{Dc}{2r}}, \quad c'_L = \sqrt{\frac{D'c'}{2r}}.$$

Then, by equation (17), page 37, we have, as the difference between the end-area solidity and the volume by the Method of Level Sections—

$$C_L = \left(\sqrt{\frac{Dc}{2r}} - \sqrt{\frac{D'c'}{2r}}\right)^2 r\,\frac{l}{6}$$

$$= \left(\frac{Dc}{2r} + \frac{D'c'}{2r} - \frac{2\sqrt{Dc}\sqrt{D'c'}}{2r}\right) r\,\frac{l}{6}$$

$$= (Dc + D'c' - 2\sqrt{DD'cc'})\,\frac{l}{12}. \qquad (a).$$

* "Easy Rules for the Measurement of Earthworks," by Ellwood Morris, C. E., pp. 150, 153, 70, etc. A tabular example is given on page 152, to illustrate this supposed necessity, in which "the grade-prism is *included* in the earlier operations and *excluded* from the later ones;" but had the grade-prism been neglected altogether, the same results would have been reached with much less labor.

It will be seen in the following pages that the above work contains some other errors in respect to accurate computation—notwithstanding its writer's deservedly high reputation as a civil engineer—which are pointed out as they come up, when connected with the present subject. The writer is induced to do this with some hesitation, because, as that

We have now to determine the difference between the *true* volume and the end-area solidity. As given by equation (15), page 38—

$$C = (c - c')\,(D - D')\,\frac{l}{12}, \qquad (b),$$

$$= (Dc + D'c' - Dc' - D'c)\,\frac{l}{12}. \qquad (c).$$

The difference *between* these two differences is the error involved in the Method of Level Sections.

Subtracting, then, equation (c) from equation (a), we have—

$$C_L - C = (cD' + c'D - 2\,\sqrt{cD'}\,\sqrt{c'D})\,\frac{l}{12}$$

$$= (\sqrt{cD'} - \sqrt{c'D})^2\,\frac{l}{12}. \qquad (d).$$

Equation (d) has always a positive value, since the square is always positive, and hence C_L is always greater than C. When, however, $\frac{c}{D} = \frac{c}{D'}$, as is always the case when the surface-slopes in each section are identical, the equation reduces to zero.

It follows that the solidity by the Method of Level Sections is always too small, except when the surface of the ground is a plane. Equation (d) is evidently independent of road-bed and slope, since terms containing w and r have disappeared in deducing it, either in the above demonstration or in the paragraphs referred to.

Let us now consider the "Method of Centre-Heights," defined in paragraph 41. The formula of correction by that method is the same as that of the Method of Level

work is recent and apparently exhaustive, its often quite erroneous statements may lead to fruitless labor; or, what is more likely, to a neglect of all more accurate methods than that of averaging end-areas.

Sections, except that the *actual* centre-height is used instead of reducing to an equivalent level centre-height. Designating the correction by that method by C_H we have—

$$C_H = (c - c')^2 r \frac{l}{6}. \qquad (e).$$

By equation (*b*), the true correction is—

$$C = (c - c')\,(D - D')\frac{l}{12}. \qquad (b).$$

These two formulæ become strictly identical only when, in two given sections, $(D - D') = 2r\,(c - c')$, which is only the case, except by accident, when the width of each section on top is equal to that due to a level section. For all ordinary surface-slopes, however, the difference is very small, as long as the surface of the ground is a plane, but $2r\,(c - c')$ is always less than $(D - D')$, causing a slight tendency to excess of solidity. When the ground is not a plane, the difference in width on top is sometimes greater than that due to level sections, and sometimes less, with no special tendency in either direction, due to the irregularity of surface.

It follows that by the Method of Centre-Heights the solidity will be sometimes too great and sometimes too small, with a slight tendency to excess.

The difference between these three methods is practically illustrated by the corrections obtained in paragraphs 43 and 45, for the two solids shown in Figs. 9 and 12, and also by the tabular examples given below, on pages 153, 155, and 156. In both instances the greater ultimate accuracy of the Method of Centre-Heights, on account of its freedom from *cumulative* error, is very clear. A few numerical examples of course do not in themselves afford sufficient basis for a general statement, but it will be always sustained by a comparative trial of any extent.* By the aid of the Diagram of Prismoidal Correction, and by using *corrections only* for purposes of comparison, such a trial can be made in an hour or two, on a mile or more of actual work. It is not necessary to determine and compare the actual *solidities* of each solid by the three different methods, but the corrections only may be used for comparison, as shown in the tabular examples below.

If these conclusions are correct, it is difficult to account for the extensive use of the "Method of Level Sections," and its indorsement by continued new publications at the present day; but it is doubtless largely due to its convenience as the basis of a tabular method—since there can be but *one* level section for a given centre-height, instead of 320 as on the Diagram of Cross-Sections—*conjoined with its seeming precision.* It is evident that, merely as a question of convenience, the Method of Centre-Heights is much more simple and concise, but, with one exception,† *all* the numerous compilers of tabular methods have used the former in some of its various modifications; apparently following, without investigation, the example of Sir John Macniell—who invented

* It may be well to add that these remarks are based in part on such a comparison over the greater part of twenty miles of heavy work. In a total of about 400,000 cubic yards, the "Method of Centre-Heights" gave an excess of a little over 400 yards, and the "Method of Level Sections" a deficiency of about 300. The error in the end-area solidity was about 6,000 yards.

† Warner's "New Theorems, Tables, and Diagrams, for the Computation of Earthworks." An exception should also be made of an ingenious and excellent system of unpublished tables, described by George B. Lake, C. E., in *Van Nostrand's Engineering Magazine*, vol. iv., p. 288, which is based on the Method of Centre-Heights in a modified form, and there may be others unknown to the writer.

or at least introduced it in the early days of railway construction, and compiled by far the most extensive series of earthwork tables which have ever been published—and of his numerous successors. The late Prof. Gillespie demonstrated its tendency to cumulative deficiency in the case of two level sections,* but failed to point out any equally simple as well as accurate method, and his conclusions seem to have since been either ignored or explicitly denied.†

Another cause, perhaps, for the general use of the method is that, on *any one given solid*, it is nearly always more correct than the "Method of Centre-Heights," and that any extended comparison involves so much labor by ordinary processes as to be out of the question. Doubtless such comparisons may have been made between the methods of "Level Sections" and "Centre-Heights" *only*, but the true volume is much more difficult to calculate, and the fact that it lies about midway between them can hardly have been determined.

To the Computation of Irregular Earthwork, all that has been said above as to the Method of Level Sections applies with equal force. The error always tends to deficiency, and is *nearly* always increased by the irregularity of the solid, though occasionally such irregularities counterbalance the tendency to deficiency. This remark is equally applicable to another still more elaborate process of computing irregular earthwork, which has been given at great length by a recent writer;‡ namely, to

* See *Journal of the Franklin Institute*, March, 1859, and "Roads and Railroads," revised edition, p. 376.

† E. g., "Easy Rules for the Measurement of Earthworks," by Ellwood Morris, C. E., pp. 150, 165, *et al.*

‡ Ibid., pp. 31, 56, etc., and Chapters II. and III., *passim.*

compute for each of the end-sections of an irregular solid a single equalizing line which shall inclose an exactly equal area, and *then* apply the prismoidal formula to the solid thus transformed. This process retains the same constant tendency to deficiency (in a slightly less degree); and in considering this question it should be remembered that the process of cross-sectioning earthwork introduces some tendency to deficiency, owing to the fact that the surface-levels are assumed to be connected by straight lines instead of curves, so that the method of computation should rather tend to counterbalance that source of error.

The process of equalization, above referred to, is in detail as follows: After the area of the original section has been computed, the lowest side-height is retained, and a new opposite side-height determined for a single uniform surface-slope, which shall include exactly the same area. Having determined the two side-heights, the solid may be computed by the formula of solidity for two-level prismoids, neglecting the centre-height and width on top, or those dimensions may be determined from the side-heights and the solid computed by the formula for three-level sections. The two processes are not identical. In the first case the surface is assumed to be a *single* hyperbolic paraboloid, and in the latter to consist of *two* hyperbolic paraboloids, of which the centre surface-line and the two side-edges are directrices. The former process is not known to have been used for this purpose, but is both simpler and more correct than the latter.*

* The following is a shorter process of making these computations than that commonly given: The area of a two-level section is $A = (h + h')\frac{w}{2} + hh'r$. If we extend the section to include the area of the "grade-prism," we have $A = hh'r$. Then, given the lower side-

We have, then, the following methods, any one of which may be used in computing irregular earthwork, to compare with each other:

First. An *exact* computation, by computing the mid-section, and applying the prismoidal formula directly.

Second. Computing the end-area solidity, and determining a correction by the "Method of Centre-Heights."

Third. Determining a similar correction by the "Method of Level Sections."

Fourth. Determining a correction as if for three-level sections, neglecting the intermediate levels, as proposed in paragraph 56.

Fifth. Equalizing the sections, by the process above given, and determining the prismoidal correction by equation (*b*), for three-level stations.

Sixth. Equalizing the sections, and computing the correction from the side-heights only, by the formula for two-level sections.*

height, h, of the original section, and its corresponding distance out, d; $h'=\frac{A}{d}$; $D=d+h'r$; and $c=\frac{A}{2D}$. The constant altitude of the grade-prism for the given road-bed has then to be deducted from c and h'.

* Two other methods are enumerated in "Easy Rules for the Measurement of Earthworks by the Prismoidal Formula," but neither of them is, strictly speaking, an application of the prismoidal formula. The first one, termed the "Method of Wedge and Prism" (p. 123), is based on a subdivision of the solid, as indicated by its name, and cannot be briefly described, as it consists of six distinct steps. The formula of error—for three-level sections—is—

$$E_w=-\frac{D-D'}{D+D'}(Dc'-D'c)\frac{l}{12}.$$

This reduces to zero when the ground is a plane; otherwise it commonly has a considerable value. There is but *one* warped solid to which

The latter formula has not been previously given, but is very similar in form to that for three-level sections, equation (*b*). Letting h, h' and $p, p' =$ the side-heights of the end-sections—

$$C = (h - p)\,(h' - p')\,r\,\frac{l}{6}\,* \qquad (f).$$

it applies correctly (one in which $D = D'$), and such a one is given as an illustration of the method (page 128 and Fig. 80); but if the direction of each surface-slope in Fig. 80 be reversed, and the *centre-heights* made identical, the method will be in error 3,333 cubic feet. To *both* solids the Method of End-Areas applies correctly.

The second method, termed the "Method of Rhomboidal Wedge and Pyramid" (page 137), is merely a complex form of the Method of Level Sections, and consists in determining level sections equal to *half* the given areas, and then turning them bottom upward upon themselves, so as to form rhomboidal end-sections. Its error is measured by equation (*d*), and hence the numerical variations on page 142 are not due to insufficient decimals, as surmised, but will be found to agree substantially with those given in the last table of this Appendix for the same solids. Of these facts, however, Mr. Morris does not seem to have been aware, since he describes this method as applicable, "*however irregular the ground may be*" (page 137); while the identical Method of Roots and Squares (equivalent to "Level Sections") is restricted to uniform surface-slopes, and to use as a test (pp. 109 and 117). It may be added that "Case 2" of this method (page 139), a modified rule for certain peculiar solids, is *not* of general application, and is only correct for the single illustrative solid given (Fig. 82 and page 144). If the centre-height of the smaller section in Fig. 82 be either increased or diminished by 10 feet, the solidity will be in error 20,000 cubic feet.

* If we compare equation (*f*) with the formula by the Method of Level Sections, we shall obtain the following equation, similar to equation (*e*) for three-level sections:

$$C_L - C = (\sqrt{hp'} - \sqrt{h'p})^2 r\,\frac{l}{6},$$

showing that the Method of Level Sections tends always to deficiency in the case of two-level sections also, as was demonstrated by the late Prof. Gillespie many years ago. (See "Roads and Railroads," revised edition, page 376.)

By comparing the formulæ for each of these methods with each other, the following propositions with regard to them may be demonstrated mathematically, but the algebraic demonstrations are omitted, as they would necessarily be of considerable length, and the subject is not so important as to require it. They are sufficiently and more concisely proved by the numerical examples given below. These propositions are:

First. *All* methods of transformation by an equalizing line, whether it be level or otherwise, tend to diminish the apparent volume of the solid, unless every part of the original surface is a plane, and the surfaces of the end-sections are "similar." In the latter case, all the preceding methods, except that of Centre-Heights, give identical results.*

Second. The Method of Level Sections gives a smaller solidity than any other process.

Third. By reducing the end-sections to a uniform surface-slope, and regarding them as three-level sections, the error is *nearly* as great as in the Method of Level Sections (always slightly less), and tends always to deficiency.

Fourth. The solidity of "equalized" sections is always greater by the formula for two-level sections than by that for three-level, but the former is less accurate than the Method of Centre-Heights.†

* The Method of Level Sections is sometimes unnecessarily limited in this respect. Thus, on page 109 of "Easy Rules for the Measurement of Earthworks," it is limited "for *exact work*" to *triangular* end-sections with a uniform ground-slope.

† The latter statement is of course empirical, and can hardly be mathematically demonstrated, not having the nature of an exact problem. By regarding the sections as "two-level" the solidity is commonly more in excess than by the Method of Centre-Heights.

Fifth. The method of paragraph 56, i. e., that of neglecting the intermediate levels in determining the prismoidal correction, has no tendency to error in either direction, except that the ground is more frequently convex than concave.

The numerical examples referred to are given in Tables I. and II., at the end of this Appendix, and in Table III., on page 152. Table I., given on page 155, applies to the sections shown in Fig. 28, and shows the effect of the different methods on three-level solids. Table II., on page 157, shows the effect of adding intermediate levels to a solid which would otherwise be bounded by plane surfaces, and hence be given correctly by all the different methods except that of Centre-Heights. Table III. is a comparison of the three most prominent methods given above on eleven exceedingly irregular solids in Morris's "Easy Rules for the Measurement of Earthworks," and includes all such examples given in that volume.

Four sections of a very ordinary description are shown in Fig. 28, with the corresponding equalized sections below. These four sections may be combined into the six different solids shown in Table I., affording types of almost every possible variety of surface with three-level sections. The comparative accuracy of the Method of Centre-Heights is very clear, but the sections were in no way specially selected, and the same comparison will be found to always hold in a trial of sufficient extent. It will be observed that the Method of Level Sections is more nearly correct for *each particular solid* in every instance.

The solid, (*a*), (*b*), in Table I. (p. 155), is bounded by plane surfaces, and hence is repeated in Fig. 29 to show the effect of adding intermediate levels. Such an intermediate level, in any section, may evidently either increase or diminish its area, giving the new sections, (*a*), (*c*), and (*b*), (*d*). Then, for a given solid, these triangular areas may be either *both additive*, giving the solid (*a*), (*b*); *both subtractive*, (*c*), (*d*); *the larger additive and the smaller subtractive*, (*a*), (*d*); or *vice versa*, (*c*), (*b*). This affords types of all common cases with but one intermediate level, except that the larger triangle might have been on the smaller section in Fig. 29, and *vice versa*. In that case the column of errors in the Method of Level Sections would have been — 35, — 211, — 182, — 60, showing that the same law still holds in that case, and hence that the error tends *invariably* to deficiency in solids similar to Fig. 29.

Table III., on page 152, illustrates the effect of the above methods on solids which are exceedingly irregular, both as three-level sections and in respect to intermediate levels; and it will be seen, by examining the column of "Errors in Corrections," that the tendency to deficiency in methods of transformation is not *absolutely* invariable, as might fairly be inferred from the two preceding tables. Without attempting an exact analysis of this question, it may be explained as follows: When the two end-sections are "similar," so that the solid is entirely bounded by plane surfaces, all methods except that of Centre-Heights give identical results. When this condition is varied from *in any way*, there is introduced a tendency to deficiency. In solids similar to Figs. 28 and 29 this tendency is invariable. In more irregular solids, the same tendency exists, but other sources of error are introduced, with no tendency in either direction, which will *occasionally* equal or outweigh it. This may be proved by the same method as that used in Fig. 29. Drawing any two dissimilar sections, and making the areas due to the intermediate levels first additive and then subtractive, the *sum* of the solidities, obtained by a method of transformation, will *always* be in deficiency, and usually they will both be in deficiency. All this is sufficiently evident from an examination of the footings of the following table:

TABLE III.,

Comparison of Different Methods of Computation on the Irregular Solids given in Morris's "Earthworks."

	CORRECTIONS IN CUBIC FEET.				ERRORS IN CORRECTIONS.		
Page from which the Solid is taken.	True Correction, Difference between End-Area and Pr. Solidity.	By Method of Level Sections.	By neglecting Int'rmediate Levels.	By equalizing Sections and regarding them as 3-level.	Method of Level Sect'ns.	Neglecting Int'rmediates.	Equalized Sect'ns.
Page 57	1,830	2,365	2,000	2,364	− 535	− 170	− 534
" 58	800	1,610	1,000	1,606	− 810	− 200	− 806
" 86	1,333	1,442	1,333	1,381	− 109	0	− 48
" 87	100	82	100	80	+ 28	0	+ 20
" 89	30	5	+ 400	4	+ 25	+ 430	+ 26
" 90	200	285	200	180	− 85	0	+ 20
" 92	900	742	500	737	+ 158	+ 400	+ 163
" "	650	812	1,000	782	− 162	− 350	− 132
" 102	72	298	292	297	− 226	− 220	− 225
" "	+ 167	48	0	25	− 215	− 167	− 192
" "	+ 34	73	58	25	− 107	− 92	− 59
Net Totals of Errors in Corrections,					−2038	− 359	−1757

These remarks apply equally to the Method of Level Sections and to the process of reducing sections to a uniform surface-slope, which give almost identical results; both tending always to deficiency, but the latter in a somewhat less degree than the former, owing to the fact that, when the surface-slopes of the equalized sections are not precisely identical, equation (*d*) has a small positive value, as in the case of all other three-level sections.*

* The close similarity in these two methods, *after* the sections have been transformed, has led Mr. Morris to class them as identical ("Easy

It may be repeated here that this subject has not been discussed on account of the intrinsic importance of the discrepancies involved, but to show that the simplest methods, which seem merely approximate, are also more nearly correct than any process of transformation, although every step in the latter has an appearance of elaborate accuracy. Such needlessly tedious methods make the use of end-area solidities almost universal, but more accurate results may be reached with an inconsiderable amount of labor, which can hardly be an objection to the most indolent.

Rules for the Measurement of Earthworks," page 56), though if they were so it is difficult to see the advantage proposed from the process of equalization. As will be seen above, however, such is not the case, and the slight discrepancies in the numerical examples on pages 59, 104, 106, etc. (always in one direction, it will be observed), are due to fundamental differences in the formulæ, and not, as is there surmised, to insufficient decimals; but the main error in that process lies in the transformation of sections into others of "*exactly* equal area."

In this connection it may be added that the point suggested on page 95, as to a correction for the position of the centre of gravity in irregular earthwork is not well taken. The altitude of the solids in Figs. 43 and 44, to which reference is made, and indeed of all other solids, should be measured perpendicularly to the plane of the end-sections.

Fig. 28.

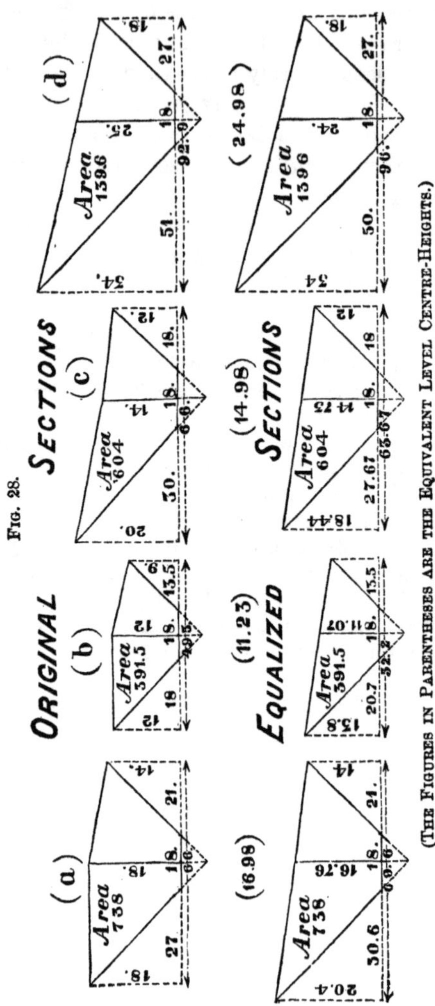

(The Figures in Parentheses are the Equivalent Level Centre-Heights.)

TABLE I.,

Showing Comparative Results from Different Methods of computing the Sections shown in Fig. 28.

Sections taken, in Fig. 28.	Solidity by averaging End-Areas.	CORRECTIONS IN CUBIC FEET.					ERRORS IN CORRECTIONS.			
		True Correction by Prismoidal Formula.	By Method of "Centre-Heights."	By Method of "Level Sections."	From Equalized Sections,		Method of "Centre-Heights."	Method of "Level Sect'ns."	From Equalized Sections,	
					regarded as "3-level" Sections.	regarded as "2-level" Sections.			regarded as "3-level."	regarded as "2-level."
(*a*), (*b*)	56,475	825	900	825	825	825	— 75	— 0	— 0	— 0
(*a*), (*c*)	67,200	0	400	100	100	98	— 400	— 100	— 100	— 98
(*a*), (*d*)	106,200	1,569	1,225	1,600	1,593	1,360	+ 344	— 31	— 24	+ 209
(*b*), (*c*)	49,875	275	100	351	350	348	+ 175	— 76	— 75	— 73
(*b*), (*d*)	88,875	4,702	4,225	4,763	4,737	4,545	+ 477	— 61	— 35	+ 157
(*c*), (*d*)	99,600	2,466	3,025	2,506	2,501	2,333	— 559	— 40	— 35	+ 133
Net totals of Errors in Corrections,							— 38	— 308	— 269	+ 328

FIG. 29.

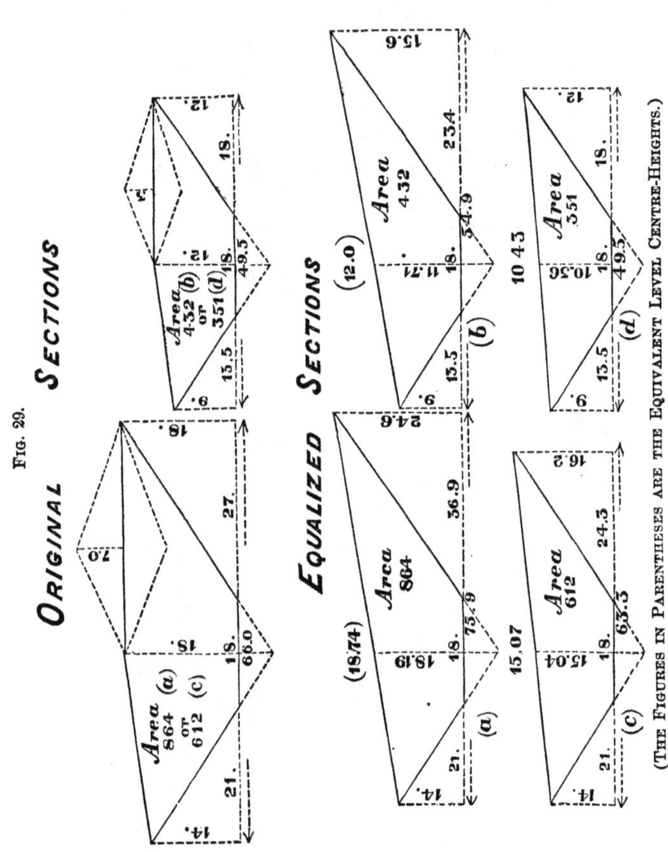

(THE FIGURES IN PARENTHESES ARE THE EQUIVALENT LEVEL CENTRE-HEIGHTS.)

TABLE II.,

Showing Comparative Results from Different Methods of computing the Sections shown in Fig. 29.

		SOLIDITY.		CORRECTIONS IN CUBIC FEET.						ERRORS IN CORRECTIONS.				
								From Equalized Sections,					Equalized Sections, regarded	
Sections taken in Fig. 29.	Middle Area.	By averaging End-Areas.	By the Prismoidal Formula.	True Correction.	By Method of Centre-Heights.	By Method of Level Sect'ns.	By Method of Par. 56, neglecting Intermediate Levels.	regard'd as 3-level Sect'ns.	regard'd as 2-level Sect'ns.	Method of Centre-Heights.	Method of Level Sect'ns.	Par. 59, neglecting Intermediates.	as 3-level Sect'ns.	as 2-level Sect'ns.
(*a*), (*b*)	631.125	64,800	63,675	1,125	900	1,136	825	1,135	1,125	+ 225	— 11	+ 300	— 10	0
(*a*), (*d*)	583.875	60,750	59,175	1,575	900	1,726	825	1,706	1,575	+ 675	— 151	+ 750	— 131	0
(*c*), (*b*)	520.875	52,200	52,125	75	900	236	825	234	75	— 825	— 161	— 750	— 159	0
(*c*), (*d*)	473.625	48,150	47,625	525	900	538	825	538	525	— 475	— 13	— 300	— 13	0
Net Totals of Errors in Corrections,										— 600	— 336	.00	— 313	00

APPENDIX B.

TABLES FOR THE CORRECTION OF "END-AREA" SOLIDITIES BY THE "METHOD OF CENTRE-HEIGHTS."

As the "Method of Centre-Heights" is the simplest of any for determining solidities by the prismoidal formula, the following tables (computed by equation (18), page 41) are added, in order to give the maximum convenience to those preferring to use it. All the quantities given in them, up to 15 and 20 feet difference of centre-heights, may also be taken off from Plate IX., as stated in paragraphs 40 and 41; but as the equation from which the tables are constructed is represented on the diagram by a single parabola, a tabulation of it is so compact as to be perhaps more convenient for practical use than to trace along the curve on the diagram with a needle-point.

With these tables, the labor required to go over the excavation of a division, at any time during the execution of the work, and take off a correction for each solid where the centre-heights differ two feet or more, is so small as to be quite inconsiderable, and it has been shown, in Appendix A, that no great objection can be made to it on the ground of inaccuracy. The exact method, however, by the Diagram of Prismoidal Correction, is but slightly more laborious, and is slightly more accurate. Their comparative merits may be judged from Appendix A; and the methods are explained in Chap. II., Art. II.

The tables are of course equally applicable to the "Method of Level Sections," after reducing each section to an equivalent level centre-height.

After taking off the corrections, care must be taken to *take the proper fractional part for fractional stations.*

For other side-slopes than those given, take corrections from the 1 to 1 table, and multiply their sum total by the ratio of slope. Thus, for 2 to 1 side-slopes, multiply by two.

NOTE.—No decimals of five-tenths are shown in the table except when the decimal is exactly five-tenths; otherwise the nearest tenth on either side is given, to facilitate taking off the correction to the nearest cubic yard.

TABLE OF CORRECTIONS BY THE "METHOD OF CENTRE-HEIGHTS."

Side-Slopes, $1\frac{1}{2}$ *to* 1.

Difference of Centre-Heights.	Correction in Cubic Yards.	Difference of Centre-Heights.	Correction in Cubic Yards.	Difference of Centre-Heights.	Correction in Cubic Yards.
0.0	0.0	**4**.0	14.8	**8**.0	59.3
.1	0.0	.1	15.6	.1	60.7
.2	0.0	.2	16.3	.2	62.3
.3	0.1	.3	17.1	.3	63.8
.4	0.1	.4	17.9	.4	65.3
.5	0.2	.5	18.7	.5	66.9
.6	0.3	.6	19.6	.6	68.4
.7	0.4	.7	20.4	.7	70.1
.8	0.6	.8	21.3	.8	71.7
.9	0.7	.9	22.2	.9	73.3
1.0	0.9	**5**.0	23.1	**9**.0	75.0
.1	1.1	.1	24.1	.1	76.7
.2	1.3	.2	25.0	.2	78.4
.3	1.6	.3	26.0	.3	80.1
.4	1.8	.4	27.0	.4	81.8
.5	2.1	.5	28.0	.5	83.6
.6	2.4	.6	29.0	.6	85.3
.7	2.7	.7	30.1	.7	87.1
.8	3.0	.8	31.1	.8	88.9
.9	3.3	.9	32.2	.9	90.7
2.0	3.7	**6**.0	33.3	**10**.0	92.6
.1	4.1	.1	34.4	.1	94.4
.2	4.4	.2	35.6	.2	96.3
.3	4.9	.3	36.7	.3	98.2
.4	5.3	.4	37.9	.4	100.1
.5	5.8	.5	39.1	.5	102.1
.6	6.3	.6	40.3	.6	104.0
.7	6.7	.7	41.6	.7	106.0
.8	7.3	.8	42.8	.8	108.0
.9	7.8	.9	44.1	.9	110.0
3.0	8.3	**7**.0	45.4	**11**.0	112.0
.1	8.9	.1	46.7	.1	114.1
.2	9.4	.2	48.0	.2	116.1
.3	10.1	.3	49.3	.3	118.2
.4	10.7	.4	50.7	.4	120.3
.5	11.3	.5	52.1	.5	122.4
.6	12.0	.6	53.4	.6	124.6
.7	12.7	.7	54.9	.7	126.7
.8	13.4	.8	56.3	.8	128.9
.9	14.1	.9	57.8	.9	131.1

TABLE OF CORRECTIONS BY THE "METHOD OF CENTRE-HEIGHTS."

Side-Slopes, $1\frac{1}{2}$ *to* 1.—(*Continued.*)

Difference of Centre-Heights.	Correction in Cubic Yards.	Difference of Centre-Heights.	Correction in Cubic Yards.	Difference of Centre-Heights.	Correction in Cubic Yards
12.0	133.3	**16.**0	237.0	**20.**0	370.4
.1	135.6	.1	240.0	.1	374.1
.2	137.8	.2	243.0	.2	377.8
.3	140.1	.3	246.0	.3	381.6
.4	142.4	.4	249.0	.4	385.3
.5	144.7	.5	252.1	.5	389.1
.6	147.0	.6	255.1	.6	392.9
.7	149.3	.7	258.2	.7	396.7
.8	151.7	.8	261.3	.8	400.6
.9	154.1	.9	264.4	.9	404.4
13.0	156.4	**17.**0	267.6	**21.**0	408.3
.1	158.9	.1	270.7	.1	412.2
.2	161.3	.2	273.9	.2	416.1
.3	163.8	.3	277.1	.3	420.1
.4	166.3	.4	280.3	.4	424.0
.5	168.7	.5	283.6	.5	428.0
.6	171.3	.6	286.8	.6	432.0
.7	173.8	.7	290.1	.7	436.0
.8	176.3	.8	293.4	.8	440.0
.9	178.9	.9	296.7	.9	444.1
14.0	181.4	**18.**0	300.0	**22.**0	448.1
.1	184.1	.1	303.3	.1	452.2
.2	186.7	.2	306.7	.2	456.3
.3	189.3	.3	310.1	.3	460.4
.4	192.0	.4	313.4	.4	464.6
.5	194.7	.5	316.9	.5	468.7
.6	197.4	.6	320.3	.6	472.9
.7	200.1	.7	323.8	.7	477.1
.8	202.8	.8	327.3	.8	481.3
.9	205.6	.9	330.7	.9	485.6
15.0	208.3	**19.**0	334.3	**23.**0	489.8
.1	211.1	.1	337.8	.1	494.1
.2	213.9	.2	341.3	.2	498.4
.3	216.7	.3	344.9	.3	502.7
.4	219.6	.4	348.4	.4	507.0
.5	222.4	.5	352.1	.5	511.3
.6	225.3	.6	355.7	.6	515.7
.7	228.2	.7	359.3	.7	520.1
.8	231.1	.8	363.0	.8	524.4
.9	234.1	.9	366.7	.9	528.9

TABLE OF CORRECTIONS BY THE METHOD OF "CENTRE-HEIGHTS."

Side-Slopes, 1 *to* 1.

Difference of Centre-Heights.	Correction in Cubic Yards.	Difference of Centre-Heights.	Correction in Cubic Yards.	Difference of Centre-Heights.	Correction in Cubic Yards.
0.0	0.0	**4**.0	9.9	**8**.0	39.6
.1	0.0	.1	10.4	.1	40.5
.2	0.0	.2	10.9	.2	41.6
.3	0.1	.3	11.4	.3	42.6
.4	0.1	.4	12.0	.4	43.6
.5	0.1	.5	12.5	.5	44.6
.6	0.2	.6	13.1	.6	45.7
.7	0.3	.7	13.6	.7	46.7
.8	0.4	.8	14.2	.8	47.8
.9	0.5	.9	14.8	.9	48.9
1.0	0.6	**5**.0	15.4	**9**.0	50.0
.1	0.7	.1	16.1	.1	51.1
.2	0.9	.2	16.7	.2	52.3
.3	1.0	.3	17.3	.3	53.4
.4	1.2	.4	18.0	.4	54.6
.5	1.4	.5	18.7	.5	55.7
.6	1.6	.6	19.4	.6	56.9
.7	1.8	.7	20.1	.7	58.1
.8	2.0	.8	20.8	.8	59.3
.9	2.2	.9	21.4	.9	60.5
2.0	2.4	**6**.0	22.2	**10**.0	61.7
.1	2.7	.1	23.0	.1	63.0
.2	3.0	.2	23.7	.2	64.2
.3	3.3	.3	24.5	.3	65.4
.4	3.6	.4	25.3	.4	66.8
.5	3.9	.5	26.1	.5	68.1
.6	4.2	.6	26.9	.6	69.4
.7	4.5	.7	27.7	.7	70.7
.8	4.8	.8	28.6	.8	72.0
.9	5.2	.9	29.4	.9	73.3
3.0	5.6	**7**.0	30.3	**11**.0	74.7
.1	5.9	.1	31.1	.1	76.1
.2	6.3	.2	32.0	.2	77.4
.3	6.7	.3	32.9	.3	78.8
.4	7.1	.4	33.8	.4	80.2
.5	7.6	.5	34.7	.5	81.6
.6	8.0	.6	35.7	.6	83.1
.7	8.4	.7	36.6	.7	84.5
.8	8.9	.8	37.6	.8	86.0
.9	9.4	.9	38.6	.9	87.4

TABLE OF CORRECTIONS BY THE "METHOD OF CENTRE-HEIGHTS."

Side-Slopes, 1 *to* 1.—(*Continued.*)

Difference of Centre-Heights.	Correction in Cubic Yards.	Difference of Centre-Heights.	Correction in Cubic Yards.	Difference of Centre-Heights.	Correction in Cubic Yds.
12.0	88.9	**16.**0	158.0	**20.**0	246.9
.1	90.4	.1	160.0	.1	249.4
.2	91.9	.2	162.0	.2	251.9
.3	93.4	.3	164.0	.3	254.4
.4	94.9	.4	166.0	.4	256.9
.5	96.4	.5	168.1	.5	259.4
.6	98.0	.6	170.1	.6	262.0
.7	99.6	.7	172.2	.7	264.5
.8	101.1	.8	174.2	.8	267.1
.9	102.7	.9	176.3	.9	269.6
13.0	104.3	**17.**0	178.4	**21.**0	272.2
.1	105.9	.1	180.5	.1	274.8
.2	107.6	.2	182.6	.2	277.4
.3	109.2	.3	184.7	.3	280.1
.4	110.8	.4	186.9	.4	282.7
.5	112.5	.5	189.0	.5	285.3
.6	114.2	.6	191.2	.6	288.0
.7	115.9	.7	193.4	.7	290.7
.8	117.6	.8	195.6	.8	293.4
.9	119.3	.9	197.8	.9	296.1
14.0	121.0	**18.**0	200.0	**22.**0	298.8
.1	122.7	.1	202.2	.1	301.4
.2	124.4	.2	204.4	.2	304.2
.3	126.2	.3	206 7	.3	307.0
.4	128.0	.4	209.0	.4	309.7
.5	129.8	.5	211.3	.5	312.5
.6	131.6	.6	213.6	.6	315.3
.7	133.4	.7	215.9	.7	318.1
.8	135.2	.8	218.2	.8	320.9
.9	137.0	.9	220.5	.9	323.7
15.0	138.9	**19.**0	222.8	**23.**0	326.6
.1	140.7	.1	225.2	.1	329.4
.2	142.6	.2	227.6	.2	332.2
.3	144.5	.3	229.9	.3	335.1
.4	146.4	.4	232.3	.4	338.0
.5	148.3	.5	234.7	.5	340.9
.6	150.2	.6	237.1	.6	343.8
.7	152.1	.7	239.6	.7	346.7
.8	154.1	.8	242.0	.8	349.6
.9	156.1	.9	244.4	.9	352.6

www.ingramcontent.com/pod-product-compliance
Lightning Source LLC
LaVergne TN
LVHW011231110826
845150LV00006B/1607

* 9 7 8 1 4 2 5 5 1 4 5 9 4 *